D1283395

Copyright © 2008
ESCO Press

All rights reserved. No part of this work covered by the copyright hereon may be reproduced in any form or by any means including but not limited to; graphic, electronic, photocopying, database storage, retrieval system, recording, taping, mechanical, or web distribution, without the prior written permission of the publisher, **ESCO Press.**

ISBN 1-930044-28-3

UPC: 651881100251

Disclaimer

This book was written as a general guide. The authors and publishers have neither liability nor can they be responsible to any person or entity for any misunderstanding, misuse, or misapplication that would cause loss or damage of any kind, including loss of rights, material, or personal injury alleged to be caused directly or indirectly by the information contained in this book.

Printed in the United States of America
76543210

ENERGY EFFICIENCY
COMFORT
CONDITIONING
PLUMBING
&
ELECTRICAL

Green Awareness

A joint effort of
Ferris State University and HVAC Excellence
Publisher: ESCO Press, 2008

Contributing Authors:

Michael J. Korcal
    Ferris State University
        Assistant Professor HVACR Department

Randy F Petit Sr
    HVAC Excellence
        Director of Program Development

Joseph R Pacella
    Ferris State University
        Assistant Professor HVACR Department

Philip Campbell
    United Association of Journeymen and Apprentices
        Training Specialist

Turner Collins
    HVAC Excellence
        Director of Technical Writing

Erik Rasmussen
    HVAC Excellence, Canada
        Canadian Education Director

Ferris State University and HVAC Excellence wish to acknowledge the contribution of the many who graciously aided in the development of this project, with special thanks to;

Steven H Allen	**United Association, HVACR**
Rodrigo Jara	**United Association, Training Department**
Richard Benkowski	**Carrier Corporation**
Tom Meyer	**Green Mechanical Council**
Scott Wills	**Scott Wills and Associates**

We would also like to acknowledge the United Association of Journeymen and Apprentices trainers for their early recognition of the need to provide their membership with Green mechanical training.

United Association

The United Association of Journeymen and Apprentices of the Plumbing and Pipe Fitting Industry of the United States and Canada (over 300,000 strong and growing) or "UA" as it is commonly known is a multi-craft union whose training department has dedicated itself to Green education within the mechanical trades of their membership which not only include the Pipe trades but extends to Heating Ventilation, Air Conditioning and Refrigeration, service and installation.

The "UA" along with its employer partners the Mechanical Contractors Association of America, the Mechanical Service Contractors of America, and the National Association of Plumbing-Heating-Cooling Contractors have developed a mobile Green training laboratory designed to supplement their 300 plus training facilities and aid in creating a greater awareness of Green Mechanical Technologies.

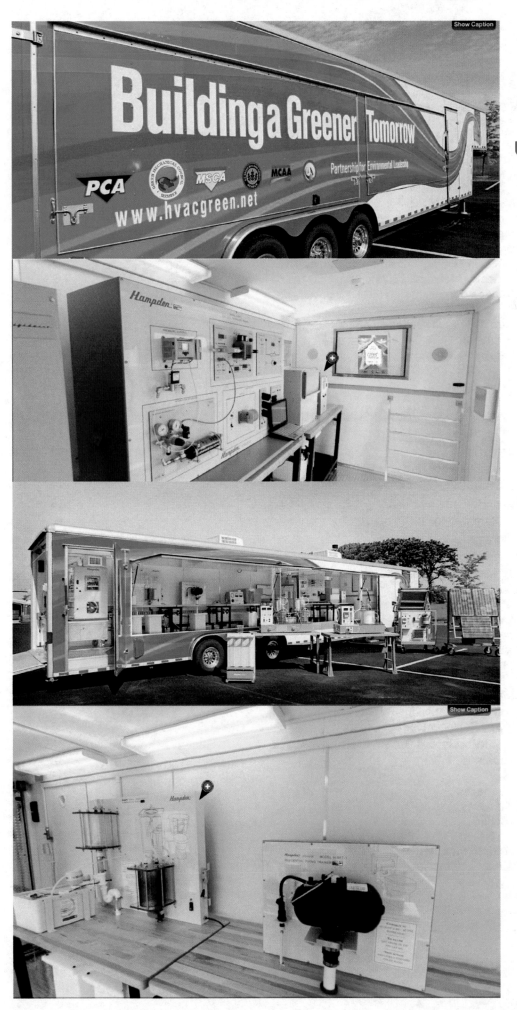

United Association's
Mobile
Green Laboratory

CONTENTS

CONTENTS

CONTENTS

Introduction

The word "Green" has become a catchall term for all things environmentally or eco friendly. The word "Green" has entered our daily lives. We read about it in the news, see it on TV and hear about it on the radio.

When we attach the word Green to our building's mechanical systems, Green takes on a more precise role in the environmental world. Green now describes energy efficiency and water conservation. For example, we would consider an electrical generating facility green, if it has extremely low emissions. Yet, if it generated power by burning non-renewable fossil fuels, it would not be considered as Green as power generated from photovoltaic (solar power), or wind turbines.

We also play a Green role. Each of us creates a personal carbon footprint. Your carbon footprint is measured by the amount of carbon dioxide that you release into the atmosphere. Carbon dioxide is recognized as a greenhouse gas and is linked to global climatic change. Burning fossil fuels such as coal, oil, and gas causes carbon gases in the forms of carbon monoxide and carbon dioxide to be released into the atmosphere.

Having an understanding of how much carbon will result from various activities, such as the burning of fossil fuels, will allow you to interpret your carbon footprint. Almost every human activity produces carbon dioxide (even breathing). The accumulation of all these activities produces large amounts of carbon. Reducing your personal carbon footprint coupled with the efforts of other people can result in a significant reduction of carbon emissions.

Worldwide, efforts are underway to reduce humanity's carbon footprint. The Kyoto Protocol established mandatory emission limitations for the reduction of greenhouse gas emissions to the signatory nations.

Green Organizations

In the United States, several organizations have been established to address the Greening of residential, commercial, and industrial buildings.

Green Mechanical Council (GreenMech) is an organization made up of mechanical system industry manufacturers, labor unions, contractor organizations, educators, students, consultants, individual contractors and others.

GreenMech is an international not-for-profit organization formed to focus on energy efficiency for the mechanical systems within our 6 million commercial buildings and 130 million housing units in the United States and Canada. According to GreenMech, "Virtually everyone is a stakeholder in energy efficiency. Each one of us must act now to maximize the efficiency of our existing mechanical systems, and specify high efficiency replacements".

"The benefits of employing Green Mechanical practices are enormous. They include the obvious cost savings, energy independence, new developments in technology, and expanded economic opportunities, while protecting the environment."

"The challenge of the task that lies ahead will require the cooperation of manufacturers, tradesman, educators and the government." www.greenmech.org.

Energy Star a joint program between the U.S. Environmental Protection Agency and the U.S. Department of Energy was designed to promote manufacturers to build products that would save energy and protect the environment. In 1992, the Energy Star program created a voluntary efficiency-labeling program for products that consumed less energy than the average or better performance as comparable models. The labeling of is designed as a way to identify the energy efficient products being sold.

Computers and computer monitors were some of the first products labeled. Since that time, major appliances, lighting systems, and more have received the label. New homes and businesses can be labeled as energy efficient by this program if constructed with Energy Star products.

The U. S. Green Building Council (USGBC) has established LEED® (Leadership in Energy and Environmental Design). This is a Green Building Rating SystemT is a nationally recognized benchmark for the design, construction, and operation of high performance green buildings. LEED establishes standards for the construction of new and the renovation of existing buildings. Buildings built under these standards are considered sustainable. The building's systems save water, and energy while maintaining indoor environmental quality.

The Green Building Initiative (GBI)

ww.thegbi.org/greenglobes operates in Canada, USA, and UK.

GBI`s Green Globe is priced modestly per self-assessment. There is an additional cost for third party verification, which includes a conditional verification at the construction documentation stage, and final verification after a site inspection is conducted.

Leadership in Energy and Environmental Design (LEED) is a Green

building rating system based on standards developed by the U.S. Green Building Council (USGBC). A building that meets or exceeds these standards is considered environmentally sustainable. A LEED certified building is a building with low energy usage and low environmental impact.

In addition to LEED certification of buildings, USGBC provides certification exams to individuals testing their knowledge of the LEED rating system. Individuals that pass the exam are recognized as "LEED AP" (Accredited Professional).

Energy Analysis and Awareness

Energy efficiency should not be confused with energy conservation.

Energy efficiency is calculated by dividing the work produced by the energy used within a process. The less energy consumed to produce the work required, the greater the energy efficiency.

It is not always easy to determine the efficiency of operating mechanical equipment. Some types of equipment vary in efficiency, based on the amount of work the equipment is performing. Most mechanical equipment works at its highest efficiency when operating at its design load.

Example: When using a simple incandescent lamp, the desired output is light, but a considerable amount of the input energy is converted to heat, and not light. The heat output is not wanted but it is a by-product of the process. An incandescent lamp is not very energy efficient.

Key Words

Energy efficiency Energy conservation
Digital
Energy recovery

Energy conservation is an effort to reduce the amount of energy needed to operate a device or process or even eliminate it.

Example: Using skylights to illuminate an area with enough light to support the function of the area. Skylights reduce and/or eliminate the input energy required by electric lighting. This is an example of energy conservation.

Methods of Increasing Efficiency:

Building maintenance: If the facility is not maintained properly, it can cause a drop in energy efficiency.

Equipment replacement: Some equipment has limited energy efficiency, even when fully maintained.

Addition of digital control: Some equipment could be more energy efficient if it were controlled more elaborately.

Energy recovery: Some equipment may have energy recovery equipment added, to increase the overall efficiency of the equipment or process.

Renewable and Sustainable Energy

Sustainable energy sources will not be depleted in a timeframe relevant to the human race. Solar, nuclear, wind, geo-thermal, and water (in the form of rain and waves) are sustainable forms of energy.

Renewable energy is a repeatable source of energy whereas the term sustainable is an energy that has no effect or impact on current or future resources. Some fuels used to produce energy are carbon based, such as fossil fuels, wood and other *bio fuels*. These energy sources impact our *carbon footprint*.

An example of a renewable energy source is *ethanol* (a type of alcohol). Ethanol can be produced through the distillation of a variety of agricultural crops and other plants. Ethanol is a *hydrocarbon* fuel. External energy is required to produce ethanol from products such as corn. Ethanol is used to operate internal combustion engines. The net result of alcohol production from corn produces slightly more energy than it takes to transform it into alcohol whereas many other crops are considerably more energy efficient. In addition to sunlight, crops for use in ethanol production require fertilizers, soil, and production management to be efficient.

Renewable and sustainable energy can reduce our energy dependence on foreign fossil fuel sources, lengthen the amount of time that fossil fuels will be available and can reduce the amount of carbon released to the atmosphere that is a contributing factor to "green house" effects.

Key Words

Renewable energy
Sustainable energy
Bio-fuel
Hydrocarbon
Ethanol
Carbon footprint

Energy Management

The term *energy management* is a very broad term and typically used to cover the whole field of energy and its use. Energy management can be divided into some more specific areas including: purchase of utility, *energy consumption* and energy *demand*, energy efficiency and energy conservation.

Computerized energy management can monitor and provide an *energy manager* with data on energy use, demand and efficiency. Using this data the manager can analyze the buildings energy requirements and project future energy needs. In many modern buildings that are equipped with sophisticated computer controlled energy management systems, the energy manager can review and control the operating temperature, humidity and other conditions within various sectors of the building. These computerized programs have a setback temperature for night and periods when the building is unoccupied. They can be programmed to shut down systems when they are not needed and start-up other systems to accommodate occupant or production requirements.

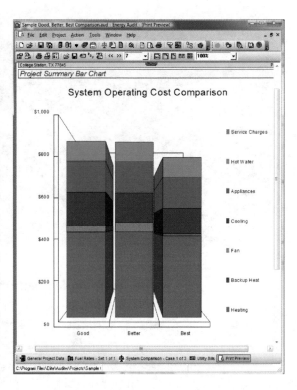

Purchase of Utilities: It is the job of the energy manager to purchase all forms of energy that must be used by a facility at the best rate. Because the utility industry is very volatile, the cost of energy must be constantly monitored so the best deal at the time can be procured. All the rules and regulations surrounding the utility rate must be understood to take advantage of the best rate. Example: A primary electrical rate will typically have lower costs for both demand and consumption but the customer is responsible for the primary transformer, which adds start up cost and owner responsibility if there is a failure. If electricity is de-regulated then a decision must be made on how to purchase the electricity (generation, transmission and distribution). Once utilities are purchased, careful analysis of each monthly bill should be done to insure proper application of rate, no mistakes and proper application of exemptions. Deregulated natural gas can be purchased very economically but must be monitored closely because of severe penalties for under or over estimating need.

Key Words

Energy Management
Energy Consumption
Demand
Energy Manager
Programming
Energy Audits
Energy Conservation Measures

Controlling Energy Demand and consumption requires knowledge of all the energy-consuming devices in the facility and how much these items consume, when they are used, and how the consumption varies with operation. Keeping equipment running only when necessary to accomplish the required work is a constant issue to keep energy demand to a minimum. Periodic *energy audits* must be done to determine if there are *energy conservation measures* that can be instituted. Equipment must be in the best-maintained condition to ensure the highest level of efficiency. Careful documentation and monitoring of all equipment can prevent deterioration of energy efficiency and increased energy consumption.

Building Information Modeling

Building Information Modeling (**BIM**) is defined as a digital computer model of the physical and functional characteristics of a facility, serving as a shared resource for exchange of information. BIM is comprised of two distinct programs.

The first program that must be run is "**Load Calculation**". The Load Calculation Program uses all building information such as construction material details, window details, lighting, other electric, personnel schedules and design weather data in order to correctly size Heating, Ventilation, and Air Conditioning and Refrigeration (**HVACR**) equipment.

The "**Building Simulation program**" is the second to be run. The stored information comes from all stakeholders in the facilities' design, construction, users, suppliers, etc. Standards are being set in order to simplify importing data from the different vendors and sources involved in the construction and operations of the facility. The Building Simulation Program uses the load calculation program above, and uses Typical Metrological Year weather data (**TMY**), along with utility and equipment data, to compute an annual energy usage and utility cost of operation.

Key Words

BIM
Load Calculation
Building Simulation
HVACR
ECM
TMY
Stake holder
Life cycle

Because BIM is a computer simulation of the facility, changes in the design of building components, heating and cooling equipment, floor plans etc. can be made before construction ever begins. This simulation predicts the effects of these Energy Conservation Measures (**ECM**). The effects of changing one item may have an effect on the construction cost overruns if any, or the operations of the facility. By having all of the information in one place, the lifecycle cost of the facility can be calculated, allowing investors to make the right choices.

The virtual digital design also aids throughout the facilities lifecycle by calculating the energy cost savings, of maintaining or replacing older systems with newer high efficient systems.

ECM What Ifs:
Example: It is determined that a building has a very inefficient chiller. The energy audit team proposes three optional ECMs to correct this problem.
1. Replace the chiller with a closed loop heat pump system coupled with geothermal,
2. Replace the chiller with a more efficient type of chiller,
3. Replace the chiller and add ice storage to take advantage of the current utility rates.

Which solution will result in the greatest energy savings and the best return on investment? The building information modeling program can be run with the different solutions and compare the cost and savings of each optional ECM.

Commercial Building Energy Consumption Survey

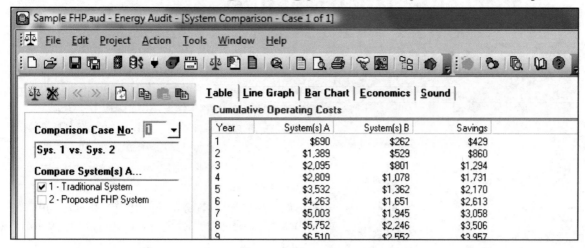

The Commercial Building Energy Consumption Survey (**CBECS**) is a national sample survey that collects information on the stock of U.S. commercial buildings; their energy related building characteristics, and their energy consumption and expenditures. The commercial sector encompasses a vast range of building types; service businesses, such as retail and wholesale stores, hotels and motels, restaurants, and hospitals, as well as certain buildings that would not be considered "commercial" in a traditional economic sense, such as public and private schools, correctional institutions, and religious and fraternal organizations. Excluded from the sector are the goods-producing industries: manufacturing, agriculture, mining, forestry and fisheries, and construction. Nearly all energy use in the commercial sector takes place in, or is associated with, the buildings that house these commercial activities. Analysis of the structures, activities, and equipment associated with different types of buildings is the clearest way to evaluate commercial sector energy use. The Commercial Buildings Energy Consumption Survey (CBECS) is a national-level sample survey of commercial buildings and their energy suppliers conducted quadrennially (previously triennially) by the Energy Information Administration (**EIA**). The 2003 CBECS was the eighth survey in the series begun in 1979. From 1979 to 1986, the survey was known as the Nonresidential Buildings Energy Consumption Survey, or **NBECS**.

Key Words

CBECS
EIA
NBECS

All of the data collected in the past can be accessed at the following web site: http://www.eia.doe.gov/emeu/consumption/index.html

The energy survey has a number of different categories but can be used to determine where a particular business compares on a national basis. Individual energy categories include items such as lighting and HVAC usage. This survey can be used as a quick reference to decide on areas of energy conservation needs or if an energy audit should be preformed to target potential ECMs. The data from this government site should be used only as a reference. In addition, data from other similar conditions should be part of the evaluation. If a business is located in a specific area of the country, it should be compared to similar business in the same area of the country so that the effect of the weather is the same.

Energy Conservation Measure

At the completion of an energy audit, there is usually a section of recommendations based on the findings of the audit. These recommendations fall into two distinct areas: Operational and Maintenance (**O&M**) issues and Energy Conservation Measures (**ECM**). The operational and maintenance issues are meant to bring existing equipment back to the original functionality and efficiency. This typically can be done by in house maintenance staff with minimal capital outlay. ECMs are more complicated to implement and usually require a professional staff and large amounts of money to institute.

Key Words

O&M
ECM
HVAC
Building Envelope

Some of the Energy Conservation Measure Areas:

Lighting - The most popular area to analyze for ECMs. The areas are checked for under lighting and over lighting. This involves changing the older, less efficient lamping with newer, more efficient lamping and fixtures.

HVAC – Heating, Ventilation, and Air Conditioning equipment usually generate the greatest number of ECMs from an energy audit. Newer equipment and controls have brought the energy efficiency ratings up to the SEER of 16+.

Building Envelope: Another popular area for ECMs is the building; such as the roof and walls. The windows can be upgraded or replaced. Air Curtains can be added to large entry ways, such as a loading dock to minimize infiltration.

Utilities: This area includes looking at specific equipment which the company or building uses, such as electric copiers for a printing company and electric golf carts for a golf course, instead of the gas type. ECMs must be discussed with the company owners and specialists familiar with the process.

Energy Information Administration

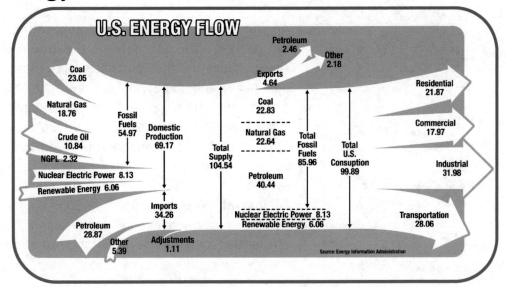

Key Words

EIA
Environment

The Energy Information Administration (EIA) was created by Congress in 1977. EIA is a statistical agency of the U.S. Department of Energy. Its function is to provide forecasts, policy data, and analyses to promote sound policymaking, efficient markets, and public understanding regarding energy. EIA develops information independently and is not subject to clearance by the other government agencies. EIA issues a range of reports on energy for other government agencies, industry, and the public.

A few of the energy reports or topics are petroleum electricity, fossil fuels, nuclear energy, alternative and renewable fuels, forecast reports, and environment reports.

More information can be found at:
http://www.eia.doe.gov

Energy Audit

An energy audit is like an accounting system for energy. On one side of the balance sheet, there is the purchase of various quantities and types of energy such as electricity, natural gas and fuel oil to name just a few. On the other side of the balance sheet is each of the energy types and how efficiently they are consumed by items such as lighting, HVAC equipment and process equipment. The results of energy audits point to possible areas for improvement in efficiencies, purchase and utilization of each energy source.

Types of Energy Audits: Energy audits are a function of time and money, the more extensive the audit, the more time it takes and the more money it costs.

The *Preliminary or Walk through Energy Audit*: This energy audit takes the least amount of time and has the lowest cost. Little or no engineering or energy calculations are done. Simple suggestions are made based on observation, such as wattage and types of lamps used, lights being left on, poor building insulation, etc. An energy balance sheet is not attempted in this form of audit. This audit usually identifies operational and maintenance (O&M) issues.

Key Words

Energy Audit
Preliminary Audit
General Audit
Technical Audit

The *General Energy* Audit expands on the preliminary audit and corrective measures, as well as costs, in detail. The utility bills for the building in the previous years (1 – 3), are reviewed and specific problem areas are specifically identified. Some engineering and calculations are run to determine potential saving from potential energy conservation measures (ECMs). The start of the energy balance sheet is done in this type of audit.

The *Technical or Engineering Energy Audit* is the most complex of the energy audits and takes the most time and money. A full analysis of a customer's purchase and utilization of all energy forms is done and is sometimes called a building performance study. Detailed energy utilization scenarios are accomplished by using building simulation software. This type of audit requires considerable engineering but can offer the greatest possible payback.

The final application of energy audit proposals are up to the building owner. Decisions are usually based on the money available but can be based on other factors. Ability to disrupt the activity or purpose of the business will be a high consideration when applying ECMs.

Energy Consumption and Demand Analysis

After conducting the Energy Audits, the next logical step is to analyze the present energy usage and energy demands of the building, industry or complex, especially the **peak demand** times or periods. Many or most electrical utilities charge the commercial and industrial consumers according to the *peak demand* of electrical energy used. These *peak demands periods* can be re-scheduled or the energy can be **spot purchased** from various sources in *time slots*. The Energy Consumption and Demand Analysis Report provides this information to the management for implementation or provide suggestions for revisions to the present methods being used. The report should include an equipment efficiency review and a report of how the electrical energy is being purchased and consumed.

Evaluation of Energy Consumption: All energy consuming equipment within a building must be analyzed. Data is collected on all equipment using nameplates, stamped information, package inserts, manufacturers specification data, etc. This data is the **baseline** information on how a piece of equipment utilizes energy. The next step is to determine the typical operating conditions of the equipment. Operating conditions can have a major impact on the energy efficiency of equipment. The desired outcome a piece of equipment must have should be compared to the equipment in place. This also has an impact on the energy efficiency of the equipment in question.

In commercial and industrial applications, preventing all of the equipment from restarting at the same time after a power outage can save on energy cost. The cost of electrical energy for commercial use is determined by demand and **kilowatt hour**. If all major energy-consuming devices are allowed to start up at the same time, the demand increase from the starting loads is used to calculate the price per kilowatt hour used for the entire month.

Key Words

Peak demand
Spot Purchased
Baseline
Kilowatt hour

Heat Load Calculation

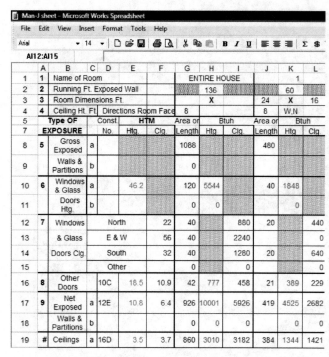

A **heat load calculation** is used to determine the **Btu** rating or the size of the heating and/or cooling system required for comfort conditioning of a structure. The heating calculation includes an evaluation of sensible **heat loss** from the structure to the colder outside air. The cooling calculation includes the amount of the **sensible heat** and **latent heat** gain, which is added to the structure along with the solar **heat gain** through the windows.

To make the calculation, the total **R-values** of the building materials are used to determine the resistance to heat transfer, along with the temperature difference between the desired indoor air temperature and the outside design air temperature. The higher the r-values of the walls, floors, ceiling, doors, and windows, the heat transfer in or out of a structure.

Key Words

Heat load calculation
Btu
Heat loss
Sensible heat
Latent heat
Heat gain
R-value

It is imperative that a "Heat Load Calculation" is performed on all buildings prior to sizing the heating and/or cooling system equipment. Over-sizing or under-sizing heating and cooling equipment can lead to excessive energy consumption, indoor air quality problems, premature equipment failure, and inadequate comfort control.

Many people ask, why not replace the old system with the same size? Changes in the structure (such as new windows, insulation, roof, doors, etc.) can change the system size requirements. In addition, the new codes for installation and the design of the higher efficiency heating and cooling systems will change the equipment size requirements for most structures. Only a correctly sized heating and cooling system will operate using our energy resources efficiently.

Life Cycle Cost Analysis

The Life Cycle Assessment or Life Cycle Cost Analysis (**LCCA**) should be understood before you specify or purchase Green products and systems.

The initial cost of a product or system, coupled with the expense of installation is frequently 10% or less of the total **Life Cycle Cost**. The remaining 90% of the cost includes the estimated lifetime power / fuel consumption and maintenance costs. Using LCCA to evaluate the cost of projects aids the builder or owner in making decisions that effect the total future cost. Cost may include an entire site complex or just a specific building system component.

If mechanical equipment is not maintained properly, the operating and repair costs will increase to more than the Life Cycle Cost Analysis prediction.

Key Words

LCCA
Life Cycle Cost

Heating-Ventilation-Air Conditioning-Refrigeration

The following section on Heating, Ventilation, Air Conditioning and Refrigeration (*HVAC/R*) identifies high performance low energy consuming products and systems currently available for comfort heating and cooling. Additional concepts and terminology related to HVAC/R are included.

Modern control systems are able to modify the operation of standard heating and cooling systems, reducing energy usage and increasing human comfort.

Key Words

HVAC/R

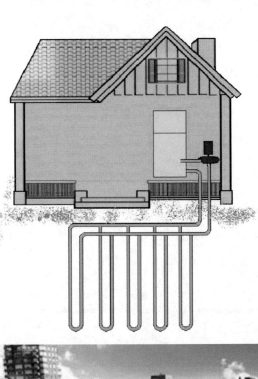

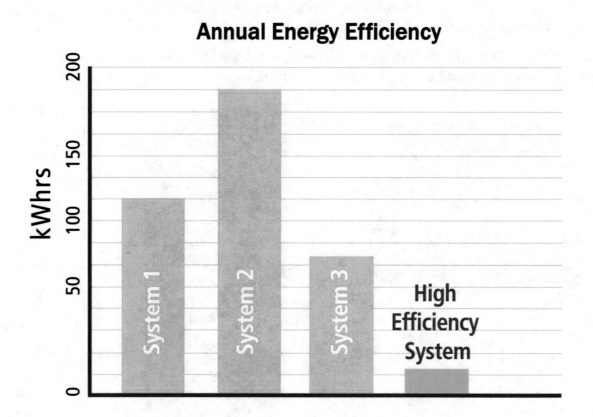

Annual Energy Efficiency

HVAC/R Energy Efficiency Ratings

♦ **Energy Efficiency Ratio**

♦ **Seasonal Energy Efficiency Ratio**

♦ **Annual Fuel Utilization Efficiency**

♦ **Heating Seasonal Performance Factor**

♦ **Coefficient of Performance**

Energy Efficiency Ratio

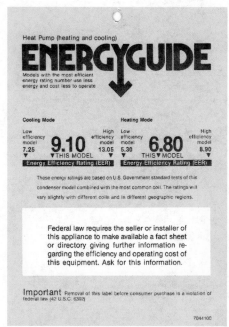

The American Heritage® Dictionary of the English Language defines **EER** as: "A measure of the relative efficiency of a heating or cooling appliance, such as an air conditioner, that is equal to the unit's output in Btu per hour divided by its consumption of energy, measured in watts." or simply put, the ratio of the cooling capacity of an air conditioning unit in Btu per hour, divided by the electrical input in watts under specified test conditions. The higher the rating number, the higher the efficiency.

The test conditions for air conditioning are 95°F. outdoor dry bulb temperature and 80°F. dry bulb/67°F. wet bulb indoor temperature.

EER ratings labels are used on window units, unitary equipment and heat pumps. The label is required by the Department of Energy (**DOE**).

Key Words

EER
DOE
SEER

Seasonal Energy Efficiency Ratio

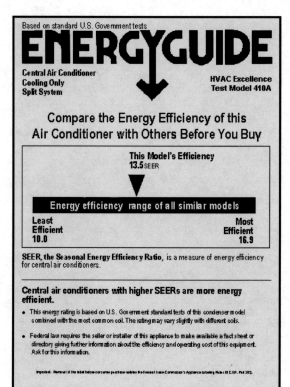

Seasonal Energy Efficiency Ratio (**SEER**) is used to measure the efficiency of a central air conditioning system. It measures how efficiently a cooling system will operate over an entire season, as the outdoor temperature changes. The more efficiently the air conditioner operates, the higher the SEER. Technically speaking, the SEER rating of the system is the ratio of the cooling output in Btu for the season, divided by the power consumption in watts per hour for the season.

Since the climate is not the same, the country is divided into five regions, with region one in the south and region five in the north. Cooling and heating hours from region four are used for calculating the SEER and HSPF ratings. For this reason, the SEER may be a little higher or lower, depending which region the equipment is to be installed.

In 1992, the minimum SEER rating was 10. On January 23, 2006, the minimum rating jumped 30% to a SEER of 13 for central air conditioning systems. Just as the EER, the higher the rating, the higher the system's efficiency.

Annual Fuel Utilization Efficiency

AFUE is a rating of how efficiently a device consumes fuel over an entire season of use. Although the testing for AFUE has been available for a long time, the term AFUE rose to importance due to the first energy crisis that gripped the United States in the 1970's. The Federal Trade Commission (*FTC*) requires that new heating equipment must display the AFUE rating on the equipment, so that consumers can compare the heating efficiencies of different equipment.

Minimum energy efficiency standards in the United States for furnaces and boilers took effect in 1992.

Key Words

AFUE
FTC

The minimum allowable AFUE rating for various systems is:

Fossil fueled forced air furnaces	78%
Fossil fueled boilers	80%
Fossil fueled steam boilers	75%

By comparing the AFUE rating of products, consumers can make informed choices and purchasing decisions. The higher the AFUE number, the greater the energy efficiency of the heating system, meaning less fuel is used to provide the required heat.

If an AFUE for a unit is 78%, this means that 78% of the energy that is stored in the fuel will be transferred into heat for the building. The 22% that is remaining leaves the building through the chimney or vent system to the outside of the building.

Due to the relatively high temperatures in the exhaust gasses, the water that is created during the combustion process remains in a vaporous state in the exhaust gases. The AFUE can be higher than the minimums listed above, and are accomplished by using equipment that removes more heat from the exhaust gasses. As the heat is recovered, the unit's combustion efficiency increases and this correlates to increases in the AFUE. At the same time, the decrease in exhaust gas temperature causes moisture to condense and precipitate out of the exhaust gasses. These devices are known as condensing appliances. It is not uncommon for condensing residential furnaces to approach AFUE levels of 96%, and condensing residential boilers can exceed 92%.

The testing procedure is found in ASHRAE Standard 103 & 124. The test has been designed by ASHRAE and involves starting the unit for a specific amount of time before measuring its combustion efficiency. The unit is then shut down for a specified time, similar to a normal off cycle for the unit. The AFUE is then calculated.

Heating Seasonal Performance Factor

Heat pump heating performance is measured by the Heating Season Performance Factor (**HSPF**), a ratio of the estimated seasonal heating output, divided by the seasonal power consumption for an average U.S. climate. Heating Seasonal Performance Factor is similar to the SEER, as it also takes into account the efficiency of the equipment for an entire heating season. The normal cycling of the system's components are taken into consideration when calculating the Heating Seasonal Performance Factor. If a system is oversized, the actual HSPF rating published may not be what the system actually achieves.

Key Word
_____

HSPF

HSPF is a measurement of how efficiently all residential, and some commercial heat pumps operate during the entire heating season. Comparing Heating Seasonal Performance Factor rating numbers will aid when making a choice in selecting a heat pump system. Whether to use the HSPF or SEER depends on the geographic location, cost of fuel, and the length of the heating or cooling season.

HSPF heat pump equipment ratings range from 7.0 to 8.0, and can be used to calculate the average coefficient of performance (COP) for an entire heating season.

Example: If the equipment has a HSPF of 7.5, the unit's COP can be calculated as:

$$COP_{Heat} = \frac{7.5000}{3.413} = 2.197$$

In other words, for every one unit of heat energy used, there will be 2.197 units of heat produced by the unit or 219.7% efficiency.

By comparison, an electric heater produces one unit of heat out for every one unit of electrical energy used.

As previously stated, HSPF is similar to SEER, EER and COP, the higher the number the greater the equipment's efficiency.

The minimum HSPF rating for heat pumps is regulated by the Department of Energy (DOE). Minimum ratings for various units as mandated by the DOE come from recommendations by ASHRAE, which are listed in ASHRAE standard 90.1-2004 table 6.8.1B).

Coefficient of Performance

Certain types of heating (heat pumps) and air conditioning (water & air-cooled chillers) products are able to provide increased system efficiencies, due to the byproducts of the mechanical energies produced during operation. This affect on the refrigeration cycle actually increases the efficiencies of the equipment.

Key Word

COP

A heat pump can use one unit of electrical energy to produce more than one unit of heat energy output, due to the effort exerted by the compressor and the refrigerant. Heat pump operating in the heating mode can achieve *COP* ratings of 4.5 or greater. This means that for every one unit of electrical energy consumed by the equipment, it will produce 4.5 units of heat energy. As a comparison, an electric heater has a COP of one, meaning that one unit of electrical energy used by the resistance heater will produce only one unit of useful heat or work.

The COP of the system will vary, according to the outdoor air temperature. The system design has to take into account the temperature of the fluid or medium with which the system is exchanging heat. The fluid could be air, ground water or re-circulated water. The manufacturer's equipment design information will list the COP as a sliding scale based upon fluid temperatures.

Coefficient of Performance (COP) is listed at 47°F. and 17°F. outdoor temperatures for air-to-air heat pumps.

The COP of geothermal heat pumps remains higher because the ground temperature does not vary as much as the air temperature. If a heat pump has a COP of 3, it can move 3 units of heat for every 1 unit of energy it consumes.

The Department of Energy (DOE), using recommendations set by the American Society of Heating, Refrigeration, and Air Conditioning Engineers (ASHRAE), mandates the COP for some systems.

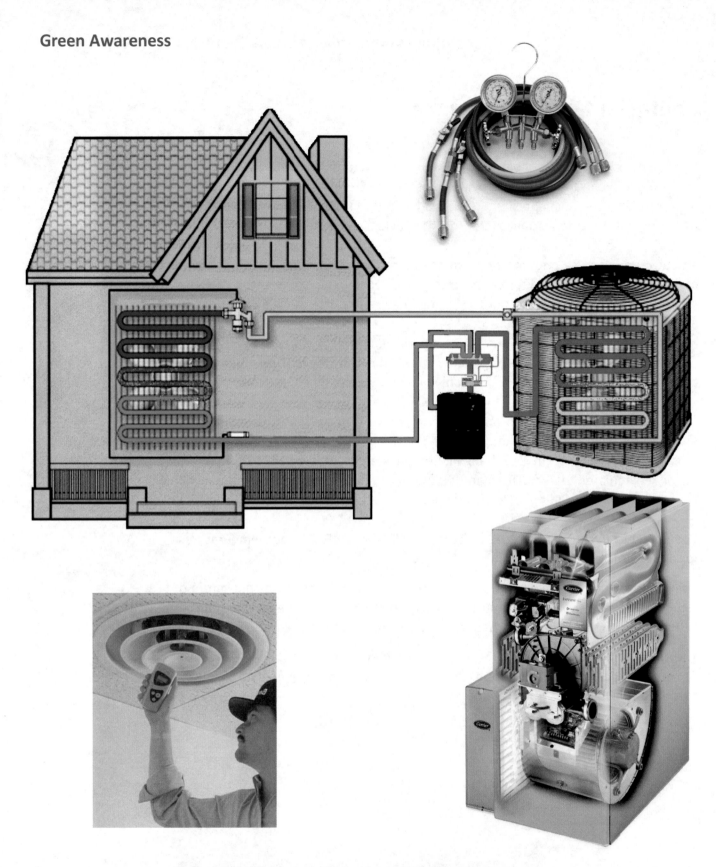

Comfort Conditioning
Temperature Control / Air Quality

Comfort Conditioning

This section identifies products and systems used for comfort heating and cooling. There are various types of systems available that can provide both comfort and efficiency, when properly selected. Additional concepts and terminology related to HVAC are also included in this section.

Before addressing HVAC, an overview of **Indoor Air Quality (IAQ)**, the quality of breathable air within a building is necessary. Occupants require air to breathe that is free from odors and hazardous materials. Hazardous materials fall into three major categories including: particulates, biological, and chemical.

Buildings constructed before the 1970s allowed far more *air infiltration* then the newer energy efficient structures. This infiltration provided enough outside air to prevent potential harm to the occupants, resulting from a lack of indoor air quality. However, as energy costs increased, construction methods and materials changed, and less air infiltration resulted. Reduced air infiltration and improper high efficient equipment sizing increased the indoor hazards, creating "*sick*" buildings.

In order to maintain an acceptable indoor air quality in tighter structures, the American Society of Heating Refrigeration and Air conditioning Engineers (**ASHRAE**) established *ventilation* standards. This process (ventilation) is measured in air changes per hour. The number of air changes per hour inside a structure, determines the required amount of outside air. Occupant levels in commercial buildings are also considered when determining the quantity of outdoor air that must be provided.

Key Words

Indoor Air Quality
Air infiltration
Ventilation
ASHRAE
Exfiltrating

Outside air that is allowed to enter a building by either mechanical or passive means is known as ventilation air, and is measured in cubic feet per minute (CFM). When air is introduced into a building, the same amount is removed in order to maintain building pressure. *Exfiltrating* is the term used to describe uncontrolled air moving out of a building.

For every action there is an equal and opposite reaction. In our quest for energy efficiency, we must be ever vigilant that our actions do not endanger human health. Airborne particulates, organisms, and vapors affect many people and can cause "Sick Building Syndrome". Air quality within buildings can be controlled to reduce and eliminate potentially hazardous substances. Reducing or eliminating airborne substances in buildings reduce medical problems.

Ventilation and Indoor Air Quality

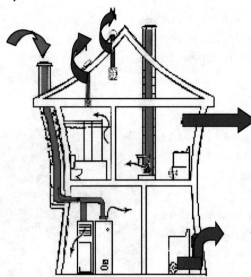

In most climates, *ventilation* air has to be either heated or cooled. Maintaining a healthy indoor air quality (*IAQ*) in commercial buildings requires that a minimum of 15 CFMs (cubic feet per minute) of outside air be brought into the building per occupant.

Over the years, many methods have been used to introduce outside air into a structure. Closely monitoring the condition of outside air is introduced into the structure can save energy. *Economizers* are used to control ventilation air by monitoring the outdoor temperature and humidity. The relative humidity in a structure should be maintained between 30 and 60 percent. High humidity will contribute to health hazards, such as mold and mildew.

Key Words

Ventilation
IAQ
Economizers
Particulates
biological organisms

There are ventilation systems available today that recover energy from the exhaust air leaving the building. These systems can save as much as 75% of the energy used to reheat the ventilation air. The type of ventilation system used depends upon the type of comfort heating and cooling system, and the outdoor climate conditions.

Contaminated outdoor air can be drawn into the structure when the heating or cooling system is operating. *Particulates* and allergens in contaminated outdoor air can lead to health problems. Both, the outdoor and indoor air can contain particulates, *biological organisms*, and chemicals. Filtering the air with a MERV 6 (minimum efficiency reporting value) or higher air filter will reduce the levels of small particles and improve the indoor environment. UV (ultraviolet light) effectively reduces or eliminates DNA-based airborne contaminants, bacteria, viruses, mold and spores that cause allergy, asthma attacks, sickness and other respiratory conditions.

Comfort Cooling Methods and Green Alternatives

The contemporary cooling methods used today are an integral part of our lifestyle, our longevity, and our social structure. Prior to air conditioning, the extreme heat of summer was often life threatening, if not deadly. People spent many hot summer evenings outdoors under the shade of a tree or protection of a porch to escape the heat of their homes. Today, nearly all new homes are constructed with a central air conditioning and heating system. All these comfort cooling conveniences have become necessities, and require enormous energy resources in both, summer and winter.

Traditional and non-traditional alternative cooling methods are being reviewed to find more efficient and cost-effective systems for residential, industrial and commercial applications. The Green alternatives and practices described in the following pages will provide options for improvements as well as some electrical energy saving methods.

Mechanical Air Conditioning

Mechanical air conditioning moves heat from a place it is not wanted, to a place where it is not objectionable, such as from the indoor living area of a structure to the outdoors.

The vast majority of comfort cooling systems used today for residential homes are air-to-air vapor compression systems. These systems, when properly sized and installed, will provide efficient comfort cooling and dehumidification. The system used for all applications can use air, water or both for the cooling mediums.

Mechanical vapor compression cooling systems are responsible for the consumption of a large portion of the total summer electrical energy production. Air conditioning equipment owners are becoming more aware of, and concerned with the operating efficiency of their systems. With rising electrical energy costs and mandated higher SEER ratings for new systems, the goal of the service personnel must be to minimize energy consumption, and optimize performance efficiency for all new and recently repaired systems.

The mechanical systems used for comfort cooling may be any or a combination of the following: Windows units, Package units, Package Terminal Air Conditioning, Central Split Systems, Mini-Split Systems, Spot Coolers, or Chillers.

Proper installation and servicing procedures for all of the mechanical cooling systems will aid in the reduction of electrical energy consumption and mechanical failures, resulting in a decrease of greenhouse gas emissions.

All of these products are manufactured with different efficiency levels. Making the correct "GREEN" choice is up to the consumer. When purchasing equipment, the consumer should make a comparison of the EER or SEER ratings.

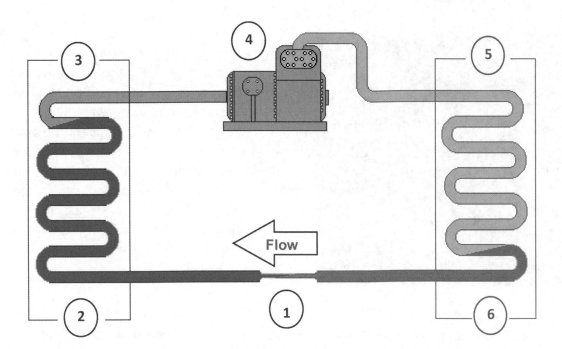

The Mechanical Vapor Compression System

Key Words

Metering device
Saturation temperature
Evaporator
Compressor
Condenser
Super heated vapor

In the vapor/ compression refrigeration cycle, liquid refrigerant at a high pressure is delivered to a ***metering device,*** **(1)**. The metering device causes a reduction in pressure, and therefore a reduction in the ***saturation*** (boiling) ***temperature*** of the refrigerant. The refrigerant then travels to the ***evaporator***, **(2)**. Heat is absorbed from the air passing over the evaporator and causes the refrigerant in the evaporator to boil from a liquid to a vapor. At the outlet of the evaporator, **(3)**, the refrigerant is now low temperature, low pressure vapor. The refrigerant then travels to the inlet of the ***compressor***, **(4)**. The refrigerant vapor is then compressed and moves to the ***condenser***, **(5)**. The refrigerant is now at a high-temperature, high-pressure ***super heated vapor***. As the refrigerant transfers its heat to the outside air, the refrigerant condenses to a liquid. At the condenser outlet, **(6)**, the refrigerant is a high-pressure liquid. The high pressure liquid refrigerant is delivered to the metering device, **(1)**, and the sequence begins again.

The Saturation temperature is the temperature at which a liquid will change to a vapor or vapor to a liquid. The saturation temperature of a given liquid will vary as the pressure changes.

Evaporative Cooling

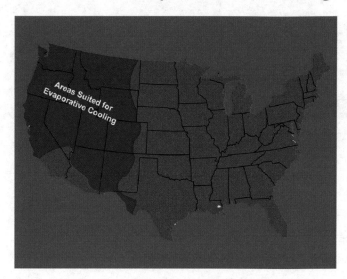

When water evaporates (changes from a liquid to a vapor), it removes 970 Btu of latent heat energy for every pound of water evaporated. As it changes, the remaining water drops in temperature. The amount of temperature drop depends on the amount of moisture in the air. The lower the relative **humidity** of the air, the more moisture will be evaporated, cooling the air. Evaporative systems work best in low humidity geographic areas where the relative humidity is in the range of 30 to 50%.

Single-Stage Swamp Cooler

One type of *evaporative cooler* is referred to as a "*Swamp Cooler*." Within the Swamp Cooler, water is misted, sprayed in the air flow or dripped over the surface of an absorbent pad. As the less humid outside air from a fan or blower flows over the pad or through the mist, the water evaporates in the air stream and the air temperature is reduced. This cool air flows into the occupied space. The cooled space must have windows or doors open to the outside for air to exit the space. As a rule of thumb, the CFM of the fan is equal to 1/2 the volume of the structure. If the air is re-circulated, the humidity will rise in the space being cooled, making the space uncomfortable for the occupants.

Key Words

Humidity
Single stage
Evaporative cooler
Swamp Cooler
Two-stage
Cascade
Heat exchanger
Misting Systems

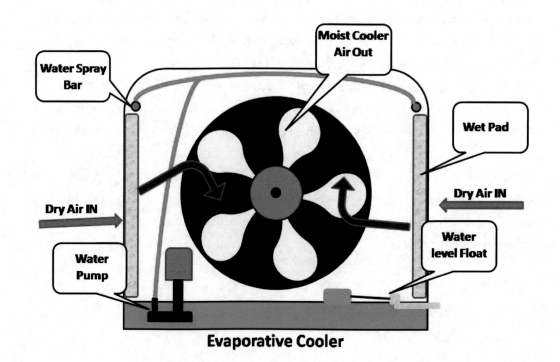

Evaporative Cooler

Two-Stage or Cascade Evaporation

Due to high humidity levels, there was a need to redesign evaporative cooling systems, so they could be used in more locations. The new *two-stage* evaporative *cascade* coolers produce less humidity compared to traditional single-stage evaporative coolers. The two-stage systems have two heat exchangers, one dry and one wet. The warm outdoor air passes through the dry heat or indirect heat exchanger first and then through the direct wet heat exchanger. Because the air is cooled a little in the first heat exchanger, it cannot absorb as much moisture in the second wet heat exchanger. The air in a single-stage system exits at approximately 75% relative humidity. The air in a two-stage system exits between 50 and 70% relative humidity.

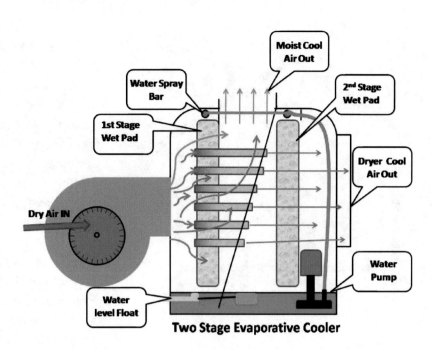

Two Stage Evaporative Cooler

The design of the **heat exchangers** and airflow through the system will be different, depending on the model. The models that use alternating dry and wet passages have better heat exchange ability. Just as the single-stage units, 100% outdoor air is used and this air must have a way to exit the building. The intake or outdoor air can be filtered, which will provide improved indoor air quality.

The systems that use a variable-speed blower to circulate cool air have a higher efficiency. As the American Society of Heating and Engineers (ASHRAE) reports indicate, two-stage evaporative coolers can reduce energy consumption by 60 to 75% compared to conventional air conditioning systems. Some can achieve a SEER nearly six times greater than conventional cooling systems.

Outdoor Misting Systems

Near waterfalls and fast running creeks, the air is always cooler. The moving water evaporates into the surrounding air, removing heat, and cooling the air. Using the same principle, a fine water mist can be sprayed into the air around an open area, reducing the air temperature and improving comfort. Commercial retailers in outlet malls are using this method on their walkways to keep their customers cool and to improve their business.

Passive Cooling Systems

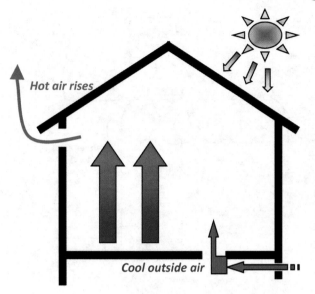

Hot air rises

Cool outside air

Air conditioning energy costs has given rise to alternate cooling methods, but these methods are not exactly new; they have been used for centuries around the world. The primary methods for transferring heat are put into practice, such as *convection, radiation and conduction.* There are many passive cooling systems available, which cost little to operate after the initial installation, and could have a tremendous impact on the total electrical energy consumption.

Green alternatives

Convection; The solar chimney uses the natural rising of warmer air through the structure to remove the heat. The warmer air is replaced by the cooler air entering at the ground level and distributed throughout the structure. This is true *passive cooling* and is very effective, as well as cost saving.

Using shade trees and open windows, as old fashioned as this may seem, can save energy and energy costs. Opening the window at night, or having shade trees located near the house can significantly help during those hot summer days and nights. These methods can be used in the moderate summer months, and some mechanical means may be needed during the higher temperature / extreme weather conditions.

Radiation; Decorative window awnings and screened-in porches are also Green alternatives that can enhance the beauty of the house, while providing protection from the *direct radiant* sunlight during the day. Window tinting will reflect the direct sun light and reduce most of the solar radiant heat from entering the structure.

Conduction; Some structures have flat roofs, with a load carrying capacity capable of holding small amounts of water, called a roof pond. The water absorbs the direct sunlight radiant heat and helps

Convection
Radiation
conduction

in keeping the occupied area beneath, cool by means of *conduction.* The water must be replaced as it evaporates. This system can be used for cooling in the summer and heating in winter. A cover is used at night when the heat is needed during the winter months for heat retention.

These alternate methods are safe, environmentally sound and can be very effective in reducing energy consumption and costs.

Solar Cooling

The sun can provide a substantial part of the energy needed for indoor cooling. Solar thermal energy can be captured by properly placed solar collector panels, which heat water or a fluid to operate the Absorption system or thermal-driven cooling process. There are absorption chiller systems available that do not operate by electricity. These systems can operate on a solar heated liquid solution instead of mechanical energy. Instead of electricity, they only need a heat source, such as the sun.

Common absorption refrigeration systems generally consist of an absorber, boiler, evaporator and condenser. The cooling system normally contains ammonia, water and hydrogen. The system operates similar to any other conventional refrigeration

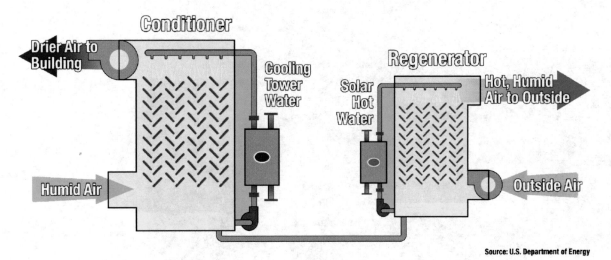

Liquid Desiccant System

Source: U.S. Department of Energy

system, except it has a boiler instead of a compressor. Absorption systems operating on solar energy can supply the cooling without the high-energy requirements for conventional air conditioning systems. They can operate with an optional heat source at night or on cloudy days, when the solar energy is not adequate.

Solar photovoltaic systems can also be used to power an air conditioning unit or a heat pump system.

Evaporative cooling is also part of the solar

cooling list of alternatives. Roof ponds, as previously discussed is an example of these energy saving evaporative cooling possibilities.

Desiccant systems are also used for solar cooling purposes by dehumidifying the air. There are two types used - liquid desiccant systems and the solid desiccant systems. After the desiccant is loaded with moisture, the liquid desiccant system uses a regenerator to drive the moisture from the desiccant material, so that it can be reused.

The solid desiccant system uses a process that moves the humid air through one section of the wheel or rotor, through the desorption process, and the dry, hot, regenerated air out the other side. The entire process is used for air dehumidification purposes. A solar hot water is used to operate the regenerator.

Key Words

Solar thermal energy
absorption refrigeration systems
Desiccant

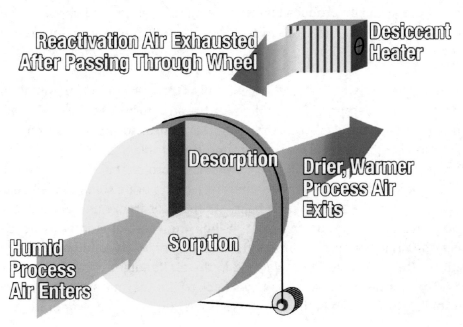

Source: U.S. Department of Energy

Thermal Storage

Thermal storage is used to store energy for heating or cooling until it is needed. Thermal storage for cooling is usually collected during the night when the demand for electricity is low and the efficiency of the refrigeration system is highest. A simple method for thermal storage cooling is to freeze water into ice and to use the melting ice for cooling during the day. To freeze one pound of water, 144 Btu of latent heat must be removed and likewise, 144 Btu must be added to melt one pound of ice. One ton of ice, melting in a 24 hour period, is equal to 12,000 Btu of cooling capacity/hour. These figures are based on freezing or melting the ice at 32°. If a brine solution is used, the temperature to freeze the solution is lowered, which will create more cooling storage capacity.

Some systems freeze ice during the cold months and then use the ice to produce cooling during the warmer months of the year. The size and cost of the storage container and the cooling load of the building are all taken into consideration in the design phase.

In the geographic locations where there are extremely cold winter temperatures, thermal storage can be produced without using refrigeration equipment. The cold outdoor temperatures can be used to make ice, which is stored and used in the summer.

In a simple system that uses ice, the ice melts as the cold water is pumped through coils to an air handler, providing cool air for the structure. Because water expands when it is frozen, the design of the system is critical to prevent damage to the refrigeration components. One way around this problem is to freeze plastic balls filled with a solution. The plastic balls are placed into a tank with a brine solution. A refrigeration system lowers the temperature of the brine solution well below the freezing temperature of the plastic balls. As the balls freeze and expand, they are free to move around inside the tank. When the thermal storage is needed for cooling, the brine solution is pumped to the air handler to condition the air in the structure. The brine solution is warmed in the air handler as it absorbs heat from the air. The warmer brine solution flows back to the thermal storage and over the plastic balls. The solution in the balls melts, lowering the brine temperature.

The system that uses ice directly is usually called a **Tank System** and the system using the plastic balls for storage is called a **Cell System**.

Key Words

Thermal storage
Tank System
Cell System

Commercial Refrigeration

This section covers commercial refrigeration as it pertains to food preservation.

Key Words

ODS

GHG

HCFCs

HFCs

There are approximately 35.000 supermarkets and a large number of cold storage distribution warehouses in the United States. The majority of the refrigeration systems in these supermarkets and cold storage distribution warehouses contain the soon to be phased-out of production, refrigerant HCFC-22, an Ozone Depleting Substance (*ODS*) and Greenhouse Gas (*GHG*). The refrigerants designed to replace hydrochlorofluorocarbons (*HCFCs)* are hydrofluorocarbons *HFCs*. HFC refrigerants do not deplete the stratospheric ozone but they are significant Greenhouse gases.

The average supermarket's refrigeration system contains 2,000 to 5,000 pounds of refrigerant, and the estimated annual refrigerant leak rate is between 25 and 35 percent. This equates to an average annual leak rate of 36,750,000 pounds of Ozone Depleting Substances and Greenhouse Gases. For this reason, it is imperative that we reduce the quantity of refrigerant required to preserve our perishables and establish better methods to contain these gases.

U.S. Environmental Protection Agency
GreenChill
Advanced Refrigeration Partnership

GreenChill is an EPA cooperative alliance to help

♦ Save the Stratospheric Ozone Layer

♦ Reduce Greenhouse Gas Emissions

♦ Improve Energy Efficiency

♦ Reduce Maintenance and Refrigerant Costs

♦ Extent Self-Life of Perishable Food Products

♦ Improve Refrigerant System Design, Operations, and Maintenance

<u>Key Words</u>

GreenChill

Refrigerant Containment Practices

Key Words

Flare fittings
Brazed
Night covers
ECM

The refrigerant leak rate can be substantially reduced by eliminating as many *flare fittings* as possible. All fittings and connections where refrigerant may be present should be *brazed*. Flare fittings should not be used in any application where refrigerant may be present, except on oil system fittings, flared pilot operated connections, and protection control devices e.g. low and high-pressure safety switches, oil pressure switches, etc. Technicians should be certified in brazing. Several organizations offer brazing certification, among them are the State Oregon, Nation Inspection Testing Certification, and the HVAC Excellence Master Specialist Brazing Certification.

A Preventative Maintenance plan designed to predict failures that may lead to refrigerant leakage could substantially reduce the annual leak rate.

Energy Conservation Measures (*ECM*)

There are numerous methods of conserving energy. Some employable measures are:

When the location is not open to the public, all open island cases should be covered with appropriate *night covers*.

The percentage of open *(without doors) medium temperature display cases, reach-in and walk in units should minimized.

All low temperature display cases, reach-in and walk in units should employ doors and should not be of the open type, with the exclusion of well and coffin type cases.

All equipment, when applicable, should employ the use of;
♦ High efficiency evaporator fan motors.
♦ High efficiency condenser fan motors.
♦ High efficiency compressor systems.
♦ Floating head pressure controls.
♦ Anti-sweat heater controls.
♦ Liquid pressure amplifiers.
♦ Mechanical subcooling.
♦ Ambient subcooling.
♦ Liquid to vapor heat exchanger.
♦ Defrost controls.
♦ Power restore sequencer or Energy Management system.

New and Replacement Equipment

Upon the replacement or purchase of new refrigeration equipment, newer technologies that require less refrigerant per ton of refrigeration than that of older designs such as Distributive or Secondary Loop systems should be considered.

Distributive systems are unlike traditional *direct expansion* refrigeration systems, which have a central refrigeration room containing multiple compressor racks. Distributed systems use multiple smaller rooftop units that connect to cases and coolers, using considerably less piping. In a distributive system, the compressors are located near the display cases, for instance, on the roof above the cases, behind a nearby wall, or even on top of or next to the case in the sales area. Thus, distributed systems typically use a smaller refrigerant charge than DX systems and hence have decreased total emissions.

Secondary loop systems use a much smaller refrigerant charge than traditional direct expansion refrigeration systems, and hence have significantly decreased total refrigerant emissions. In secondary loop systems, two liquids are used. The first is a cold fluid, often a brine solution, which is pumped throughout the store to remove heat from the display equipment. The second is a refrigerant used to cool the cold fluid that travels around the equipment. Secondary loop systems can operate with two to four separate loops and chiller systems depending on the temperatures needed for the display cases. Their average leak rate is between 2 and 15 percent per annum.

Key Words

Distributive systems
Direct expansion
Secondary loop systems

Comfort Heating Methods and Green Alternatives

The purpose of comfort heating systems is to produce, maintain, and supply warmth creating personal space comfort at work, home, and play. In our homes, we have forced air furnaces, boilers, and radiant heating systems. At work, we may have a combination of several systems radiant panels, forced air, boilers, and unit heaters. The means of producing and distributing heat are numerous. In this section, we will detail a few of the available comfort heating systems, their applications, and efficiencies.

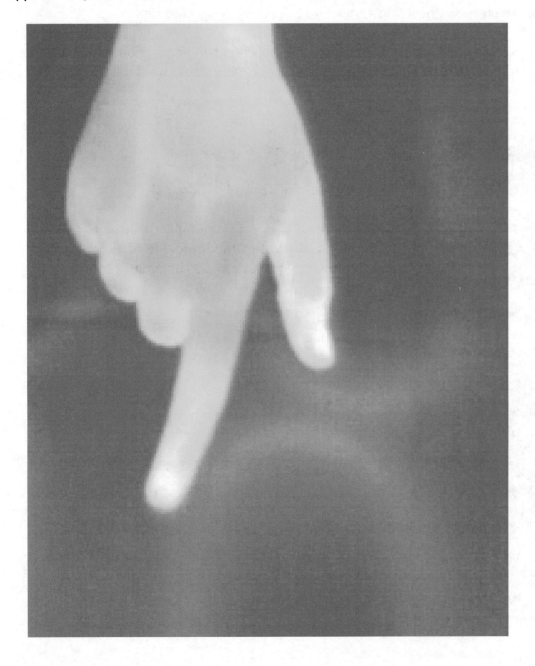

Key Words

Combustion Analysis

Combustion analysis
Carbon dioxide
Flue gas

Combustion analysis is a quintessential task that must performed to insure that combustion systems are operating to the design specification. The process of measuring the *flue gas* constituents is used to quantify the percentage of carbon that is being converted to *carbon dioxide* through the combustion process. This is done by utilizing electronic combustion testing equipment with electrochemical sensor technology or equal.

Proper testing and adjustment procedures are necessary to maximize efficiency. A furnace or boiler that is operating with a co^2 that is 1 % lower than the units design parameters may be losing as much as 5 % of its energy efficiency.

Along with a combustion analysis, numerous other verification tests must be performed to maximize a heating systems efficiency and longevity.

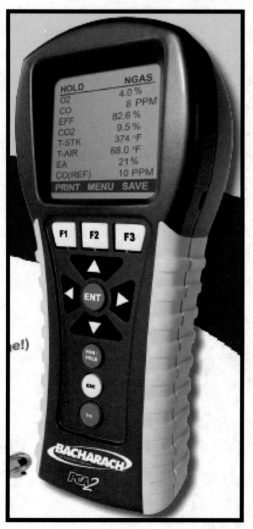

Tests such as
- Temperature rise
- Air flow (external static pressure)
- Fuel pressure
- Smoke density (for fuel oil burners)

Taking the time to perform a simple battery of verification tests can give insight to the useful life of existing systems and the potential for saving upon upgrade.

Forced Air Heating

Key Words

Forced air
Heating medium
Mid efficiency
British Thermal Unit

Forced air heating systems use "air" as their heating medium. The heat to warm the air is produced by a variety of different energy sources including electricity, natural gas, propane gas, and fuel oil.

Heating medium: is defined as any solid or fluid such as water, steam, or air that is used to convey heat from a heat source, either directly or through a suitable heating device, to a substance or space being heated.

Fuel fired heating systems are available in both mid and high efficiency models. Our focus will be on the high efficiency models. Both mid and high efficiency gas combustion systems burn the fuel identically. It is the amount of heat that is extracted that makes the difference.

The non-condensing or mid efficiency gas furnace produces heat by combusting fuel inside a heat exchanger. Air is circulated over the surfaces of the heat exchanger where it absorbs heat. Heated air is then returned to the living space.

Mid efficiency gas furnaces have vent temperature in excess of 325°F to prevent condensation within the vent system.

Condensing Furnace

The combustion of *hydrocarbon* fuels produces water vapor and carbon dioxide. Unlike the mid efficiency furnace that maintains a vent temperature of 325 °F to prevent condensation, a high efficiency *condensing* furnace operates with a vent temperature of less than 140°F.

For water vapor produced by the combustion process to remain in a gaseous state, the vapor must retain energy (heat). By cooling the vapor below its dew point, (the temperature at which a change of state occurs where the vapor becomes water again) the retained heat energy is released.

Condensing furnaces utilize an additional heat exchanger to cool and extract enough heat from the flue gases to cause the water vapor to condense.

In order for water vapor to condense, the vapor must release 970 BTU latent heat energy per pound of vapor condensate. This release of heat adds the additional efficiency to a high efficiency furnace.

High efficiency condensing furnaces are available in all the major fuel types; Natural gas, propane gas and fuel oil.

Key Words

Hydrocarbon
Condensing
Natural gas
Fuel oil
Propane

Condensing forced air fuel oil

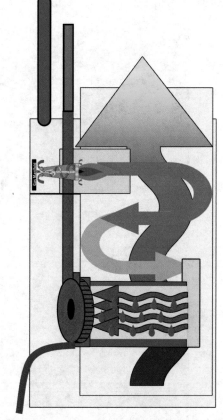

Condensing forced air natural gas or propane

Modulating Furnaces

Key Words

Heat loss
Modulating
Heat output
Blower motor

Advancements in control technologies and the demand for greater comfort have led to the development of modulating heating equipment... **Modulating** heating equipment has the advantage of being able to vary the **heat output** of the furnace to match the **heat loss** of a structure.

The modulating furnace utilizes the advanced control technologies to vary the fuel input and **blower motor** speeds to adjust the heat output of the appliance to match the heat loss of the structure. The modulation of the fuel input and blower motor speeds allows the furnace operate longer, quieter and at a higher efficiency.

Through modulation, the furnace maintains a more even temperature throughout the home, increasing the heating system efficiency and the comfort level of the occupants.

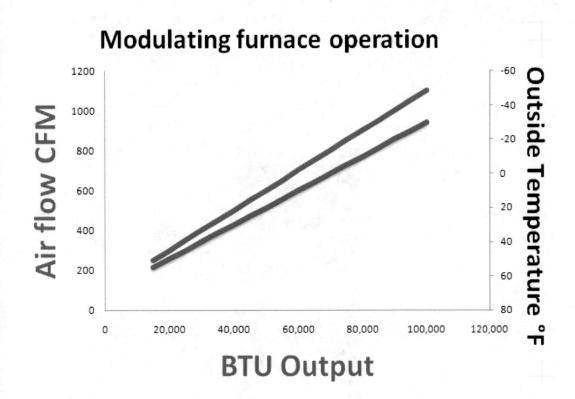

Condensation
Latent heat

Condensing Boiler

Like condensing furnaces condensing boilers, extract heat from the flue gases to the point of *condensation*, releasing the *latent heat* stored in the water vapor produced by the combustion process. This condensing oil fired boiler routinely reaches efficiency's of 92%.

Condensing boilers are available in all major fuel types, natural gas, fuel oil and propane gas.

Boiler systems can be quite elaborate in scope and versatility. Water's ability to transfer heat, makes operating multiple heat transfer systems from one high efficiency boiler very easy. One High efficiency boiler can be connected to multiple systems at one time, e.g. as indirect water heating storage tanks, fan coils, radiant floor systems, radiant panels, and traditional cast iron or baseboard radiators.

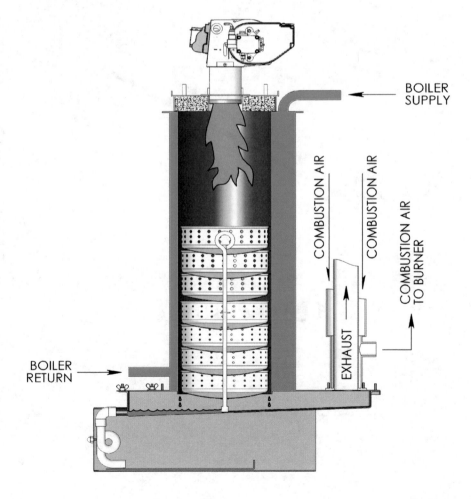

Instantaneous Boiler

Instantaneous boiler systems heat only the water required to satisfy the demand, and are sized accordingly. The reduction in **standby loss**, and the ability to locate a smaller **tank-less** boilers closer to the load, results in measureable savings. The installation of multiple boilers allows greater loads to be satisfied. Both fixed rate and modulating input boilers are available, adding even greater versatility to the installation of tank-less boilers.

Instantaneous Boiler
Standby loss
Tank-less

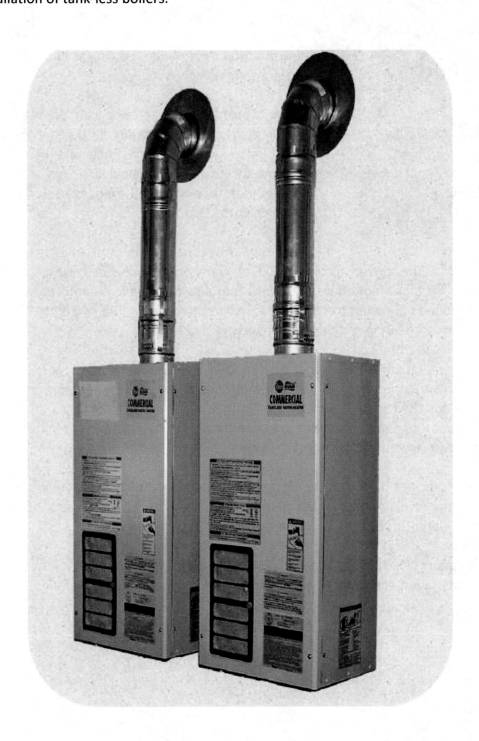

Solar Water Comfort Heating

About three decades ago, a movement began to look at the use of *solar energy systems* to heat our buildings, water and homes. Now that we are experiencing a most serious energy shortage and record high-energy costs, solar water comfort heating is at the forefront of alternative water heating methods. With the extremely invaluable initiations of those first solar energy systems, much useful information, new ideas and methods have emerged.

Solar water comfort heating can be used for heating structures or homes by transferring the solar heated water from the collectors (often roof mounted) to heat exchangers for comfort heating, or to water tank exchangers for water heating purposes.

Key Words

Solar energy
Passive
Active
Glycol

Solar water heating system must have a collector mounted in the correct position and at the proper angle in relation to sun exposure for maximum efficiency. The water also must have an antifreeze agent (glycol) added to prevent freezing in the cold temperatures. The two types of solar water comfort heating systems are the *passive* and *active*.

Passive systems do not use a circulating pump, but depend upon gravity or a siphon effect to circulate the water through the system. These systems have the collectors on the roof or mounted above the indoor heat exchanger. As the water circulates through the heat exchanger, the blower motor distributes the heat into the home or structure for comfort heating.

Active solar water heating systems often use collectors mounted on the ground or at the same level as the heat exchanger. These systems rely on a water pump to circulate water through the heat exchanger.

As with many other energy alternatives, maintenance must be done to insure that proper operation and efficiency is sustained.

Solar Air Heating

Absorber
Solar Plate
Photovoltaic

Solar air heating can be achieved through the use of a perforated-plate solar absorber system. This system is constructed using metal siding panels, black in color, that are perforated to allow air flow through them. The panels are mounted with airspace between the panel, and the outside wall of the structure. The panels are heated by the sun, and the air is heated by its contact with the heated metal. The heated air is then ducted into the structures ventilation system where it offsets the required energy to heat the structures interior. Panel location and exposure to the sun is critical to system efficiency.

A perforated plate solar absorber system can be combined with *photovoltaic* panels used to produce electricity. This combination system utilizes the same area as each individual system, but supplies two types of energy, electricity and heat. The perforated plate system also helps cool the photovoltaic panels maximizing their efficiency.

Waste heat
Heat exchanger
Condenser

Waste-Heat Recovery

Waste-Heat Recovery systems can be adapted to both residential and commercial, air conditioning and refrigeration equipment that remove unwanted heat from a space and expels it to an area where it is not objectionable, usually outdoors. The expelled heat is wasted (unused) energy. With the addition of a refrigerant to water *heat exchanger* (a device that moves heat energy from one fluid to another while maintaining a complete fluid separation) and circulating pump to an air conditioning or refrigeration system the heat can be captured and used for domestic water heating.

Another method for waste heat recovery in commercial applications is to direct warm air from the refrigeration equipment's *condenser* (the part of a refrigeration system that removes heat from refrigerant), back into the building to assist in comfort heating. This application is ideal for commercial businesses that have food storage equipment.

In some applications, waste-heat recovery can provide most of the domestic water heating needs during the summer months. In commercial applications, waste-heat recovery systems, connected to food storage equipment can provide the water heating needs all year. Based on a 60° rise in temperature, the amount of water heating can be as high as 10 gallons per hour from a 12,000 BTU (1 ton) system, running one hour. The average savings for a family of four can amount to 3500 kWh per year.

Radiant Panel Systems

Radiant panel systems are being incorporated into many new construction projects and retrofitted into many existing structures. The even heating and ease of *zoning* combine to make a versatile heating system.

Radiant panels can also be used for cooling applications. Humidity control is vital to prevent condensation of moisture on radiant *cooling panels*.

Radiant panel systems are versatile in application and design and available in many configurations such as, floor, ceiling and wall panels. The ability to zone radiant panel systems, allows users the economy of only heating the zones in use, reducing the wasted energy used to heat an unoccupied space.

Radiant panels operate on the principle that radiant energy travels through the air without transferring heat to the air. Radiant heat only transfers it energy when it contacts an object, such as a wall, desk, chair, person etc. This warm to-touch-sensation created by radiant energy provides a very high comfort level for the occupants. This also allows radiant systems to operate at lower temperatures then conventional warm air heating systems. The installation and applications for radiant systems are not limited to one fuel type. Radiant systems are available in heated fluid circulation and electric resistance types. Using radiant panel systems are a viable Green Alternative no matter what type of energy you use.

Thermal Mass

Thermal mass of a structure is a direct measure of its ability to store *thermal energy*. How much energy stored is proportional to the thermal capacity (thermal storage) of the building materials used in construction. *Dense materials* like concrete and brick have a greater thermal storage capacity in comparisons to wood and vinyl.

The Earth is a huge thermal mass; the oceans are the largest single thermal mass as water has the one of the highest thermal retention of all natural substances.

Incorporating thermal mass and storage in the design and construction of new structures can reduce a building's cooling, heating, fan energy consumption, and delay the electrical requirement to of peak times.

Simple in its construction, one thermal storage system is the *Trombe Wall*. The basic Trombe wall consists of a concrete wall, a shade and a glass outer layer spaced a few inches from the concrete. The location and angle of exposure to the sun is critical to a Trombe wall's effectiveness. By design, a Trombe wall is exposed to the low angle radiation from the sun during the winter months and shaded from the sun in the summer months. Proper shading for a Trombe wall during the summer months is critical to prevent overheating of the structure and adding to the cooling load.

Key Words

Thermal mass

Thermal energy

Dense materials

Trombe wall

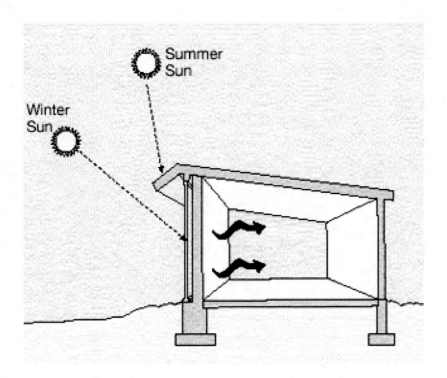

Key Words

Optimized Steam Systems

Optimized steam
Balancing

Optimized steam systems are generally used in commercial and light commercial applications. This system can return real value after installation, but the system requires proper steam system design, maintenance, balancing, and repairs. The *balancing* of an optimized steam system process refers to the continuous matching of the steam supply to the demand at a given time, which is a very involved process. There are environmental regulations and steam operator certifications that must also be considered for optimized steam systems. Steam, however, is another alternative for providing heat for business places and help in reaching our Green goals.

Steam Trap Management

Steam is an efficient heat transfer medium that is easily distributed throughout buildings and facilities. During the transmission process, the steam must be kept hot to reduce **condensation**. Some condensation is inevitable due to heat loss in the piping system. It is important to remove the condensate from the steam piping to prevent water hammering within the piping system. **Water hammers** are caused by steam velocity, which creates waves of un-drained condensates.

Thermal shock within a piping system can also occur if high temperature steam comes in contact with condensate of a lower temperature. **Steam traps** are designed to drain condensate from the piping system. The removed condensate should be returned to the boiler while it is hot, thus requiring less energy to reconvert the condensate to steam.

Many steam systems have traps that are malfunctioning. Malfunctioning traps may not remove condensate or can be stuck open bleeding off volumes of hot steam wasting energy. Therefore, it is imperative for steam system operation, and efficiency that steam trap maintenance be performed at regular intervals. Many companies have developed software programs specifically designed for tracking and co-coordinating steam trap maintenance schedules, improving steam trap management and system efficiency.

© Armstrong International, Inc.

Comfort Heating / Cooling Combination Systems

There are many Combination Systems that include cooling and heating. This section will cover heat pumps (*air to air*), geo-thermal systems (**water or ground source *heat pumps***), packaged terminal air conditioners (***PTACs***), central air conditioning with gas, oil, or electric heat, and dual energy source heat pump systems.

Key Words

Heat pumps
  Air-to-air
  Water source
  Ground source
PTAC
Geothermal
COP

Nearly everyone is familiar with the air-to-air heat pump systems. These systems are very efficient and dependable for heating and cooling. They are available in both, Packaged and Split Systems. The Package System has everything located outdoors, in one enclosure, except the ducts and control thermostat.

The Split System has an outdoor unit that houses components such as the compressor, reversing valve, and outdoor fan/outdoor coil section; the indoor unit houses the air handler, the indoor coil, and auxiliary heating unit. Historically R-22, an HCFC with ozone depleting characteristics, has been the refrigerant of choice for these systems. R-22 is being phased-out and new generation of greener or more environmentally friendly refrigerants HFC's. R-410A, an HFC, is the choice of most manufacturers for use in many new and replacement unit applications. Some of the newer heat pumps systems have SEER ratings of 16 or greater.

As energy costs rise, **Geothermal** systems become more popular. They can have Co-Efficient of Performance (**COP**) ratings of nearly 4 to 1. (Coefficient of performance (COP) - a term used to measure the efficiency of a heat pump system. It is defined as the heat output of a heat pump or electric elements, divided by the heating value of power consumed in watts at standard test conditions of 17º and 47º F. Electric resistance heating elements are rated at 1 to 1.) These systems extract the heat from the ground or from a water source. They are used for both, heating and for cooling.

PTACs or package terminal air conditioners are unitary systems used for both heating and cooling. Heat is usually provided by a resistance electric heater in the unit. Some PTACs use air conditioner technology while others contain a small heat pump. PTACs are commonly used in hotels and motels to provide guests with comfort conditioning, individual to their needs. One of the great advantages of a PTAC is that it can supply heat or cooling for one area or room, without conditioning the entire structure: Another Energy Saving, Green alternative.

Early in the 1970's and then again in 1988, variable speed systems were introduced into the residential market. Due to their cost, they did not become popular at that time. Now with rising energy cost, many manufacturers are reintroducing the systems into the market place.

Geothermal System

Most refrigeration systems use air for the exchange of heat. **Geothermal** is a special type of refrigeration unit that utilizes the earth as both a heat sink and a heat source. Refrigeration systems are composed of four basic components: compressor, metering device, condenser, and evaporator. The basic refrigeration principle relies on the ability of a material to change state (liquid to gas) and in doing so, help transfer heat. The evaporator absorbs heat to the refrigerant making the surrounding area colder. The condenser transfers the heat to the surrounding area making it warmer.

The ground is a **thermal mass**. It can be used to absorb unwanted heat for comfort cooling or transfer heat to homes and businesses during colder weather. Systems that use the ground as a heat sink (thermal mass) are called geothermal systems. Water in wells, lakes, ponds, etc. may also be used as a thermal mass in geothermal systems.

Systems that transfer heat to or from the ground or water are called geothermal Heat Pumps. Heat pumps use a reversing valve to control the flow of refrigerant switching the operation of the heat exchangers. There are many varying designs of geothermal heat pumps.

Geothermal units are among the most efficient heating and cooling systems currently available with the direct expansion the most efficient. Compared to a comparable gas fired heating system, the cost of operation is reduced by 66%. One major concern is the initial cost of installing a geothermal system. The installation cost can be 5 times the cost of traditional heating and cooling systems. These systems require qualified installation and set-up personnel. These systems may be coupled with solar panels to provide the required electrical energy source.

Systems
Single Pass
Single pass (Pump and dump): Water is taken from the ground and passes through the heat pump once. It is then dumped back to the environment. An inexpensive source of water is necessary. The only impact on the water is a change in temperature of the water, warmer in the summer and cooler in the winter.

Horizontal Loop

Fluid circulates in a loop installed horizontally in the ground. The length of the loop is determined by the soil type, depth, and the system BTU capacity (approximately 400 feet per 12,000 BTUs). To move heat between the heat pump and the ground, the fluid is continually circulated in the closed loop. The system requires enough open land to install the length of horizontal loop needed to transfer the heat.

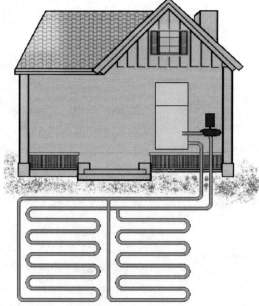

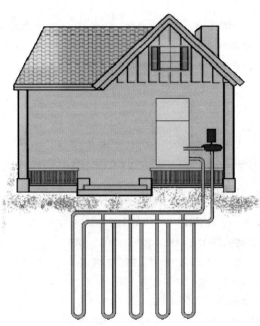

Vertical Loop

The fluid circulates in a loop installed vertically in the ground. The layout of this system is similar to a common well. Fluid continually circulates as in the horizontal loop system. The vertical loop requires less ground surface area and length compared to the horizontal loop (approximately 250 feet of vertical loop per 12,000 BTUs). Several properly spaced wells may be used to meet the needs of the system's capacity.

Direct Expansion

In the Direct Expansion (DX) system, the refrigerant tubing is in direct contact with the ground. With the possibility of a refrigerant leak, a specially developed heat exchange system along with environmentally friendly refrigerant must be used to prevent ground contamination. Direct Expansion systems have the advantage of not needing a secondary heat exchanger between the ground and the refrigerant.

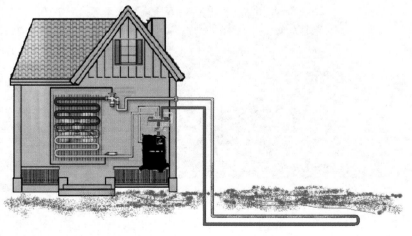

Air To Air Heat Pumps

Air to air *heat pump* systems are similar to geothermal systems in principle. Air to air systems use the air as their thermal mass.

Cooling Refrigerant in the system evaporates within the indoor coil when it absorbs heat from the return air passing over the coil. Once heat is extracted from the air and the air enters the occupied space it is called the supply air. The refrigerant having absorbed heat then travels to the outdoor coil were it gives up its heat to the outdoor air.

A heat pump is very similar to an air conditioner. The difference is in the heating mode. A heat pump has a ***reversing valve,*** which reverses the refrigerant flow in the coils and refrigerant lines. Refrigerant in the system evaporates in the ***outdoor coil*** by absorbing heat from air passing over the coil. The refrigerant then travels to the ***indoor coil*** were it gives up its heat to the indoor air.

Refrigerant is a fluid that absorbs heat when it evaporates (changes from a liquid to a vapor) and releases heat when it condensers (changes from a vapor to a liquid).

Key Words

Air to air *heat pump*
Reversing valve
Outdoor coil
Indoor coil
Dehumidification
Strip heat
Air Source

In order to provide proper ***dehumidification*** during the cooling season, heat pumps are sized based on cooling requirements. For this reason, the capacity of heat pumps may not be sufficient to satisfy the entire heating requirement. Where required, electric resistance strips or a gas burning system provides auxiliary heat.

Air to air Heat Pumps are up to three times more efficient than electric resistance heat. At any outdoor temperature, the coefficient of performance (COP) for electric heat is always 1 to 1. At an outdoor temperature of 47°F most air to air heat pumps have a COP of greater than 3 to 1 and at 10°F heat pumps have around a 1 to 1 COP. Heat pumps when properly installed and sized can reduce energy consumption and our carbon footprint.

In some areas where fossil fuels are not readily available, local code requires that homes be heated by a heat pump instead of electric resistance heat.

Packaged Terminal Air Conditioners (PTAC)

Package Terminal Air Conditioner (*PTAC*) - is an air conditioning system in which all components are in a single cabinet (*unitary*). They provide total zone control in residential and commercial applications such as hotels, motels and room additions. This system of total zone control reduces energy consumption by providing comfort heating and cooling to occupied areas on demand.

PTACs use a vapor compression system to move heat from a place where it is not wanted to a place where it is less objectionable and generally provide heat from *electric resistance strips* or a gas burning system.

Some PTACs may contain a heat pump and would operate the same as any air-to-air heat pump. PTACs that contain a standard air conditioning vapor compression system require electric resistance strips or a gas burning system to provide heat.

Key Words

PTAC
Unitary
Electric resistance strips

Mini-Split Systems

Mini split systems have many advantages over central air conditioning systems or window units. In buildings that were not designed with air ducts or make duct installation difficult, mini-split systems have the advantage of not requiring ducts. They may be flush-mounted into a drop ceiling, hung from the ceiling, or mounted on the wall. The unit is made in two primary sections; the outdoor unit, and the *indoor air handler/*evaporator. Connection of the refrigerant lines and electrical lines between the two components are made through a 2 to 3 inch hole in the wall. The condensate drains also runs from the air handlers, through the same hole in the wall.

Mini -splits can have up to four indoor air handling units (for four zones or rooms) connected to one outdoor unit. The number depends on how much heating or cooling is required for the building, or for each zone. The capacity of the zones can be different, as long as they match the total capacity of the outdoor section. Each zone must be carefully sized so that the system will not short cycle, which wastes energy and does not provide proper temperature or humidity control. To maintain proper temperatures and humidity in a zone, a load calculation should be performed for proper load sizing.

Key Words

Mini-split
Indoor air-handler

The primary disadvantage of mini-splits is their cost. These systems normally cost about $1,500-$2,000 per ton (12,000 Btu per hour) of cooling capacity. This is more than central systems, and maybe twice as much as window units of similar capacity. However, the benefits often outweigh the cost, depending on the efficiency of the system. There is normally a room thermostat for each zone to control the room temperature. There are systems available that are cooling only; some have cooling and electric heat; some are heat pump systems. Efficiencies as high as 21 SEER are available.

Electrical

Electricity is commonly being produced by burning fossil fuels. These fuels are expensive, non-sustainable, and non-renewable. They emit greenhouse gases and add pollutants to the atmosphere.

Various alternative methods for electrical generation are available such as, solar, wind, tidal, hydroelectric and nuclear.

The ever-growing demand for electricity along with the onset of global warming (from burning fossil fuels) has brought the necessity for better alternatives and more sustainable energy production resources. New technologies are providing many of these alternatives, which can produce electricity in a clean, efficient and sustainable manner.

In addition to the electric generation, it is important that consumers specify and use energy efficient appliances, lighting, and other equipment. By using the existing clean and new alternatives and methods described in this book, we can reduce the electrical demand, energy consumption, and our carbon footprint.

Electrical Power

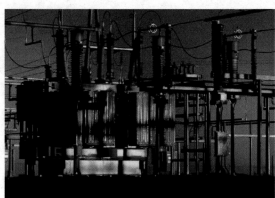

The cost for electricity is based on **Watts** (Power) consumed over a period of time. If a 60-watt light bulb is energized for 5 hours, the power consumption is 300 watt-hours. The cost of energy consumption is billed in kilowatt-hours. It takes 1000 watts to equal 1 kilowatt.

There is a mathematical relationship between **Voltage**, Amperage and Watts.

Voltage X Amperage = Watts (Power)

By understanding the electrical mathematical formula and relationship, we can gain an insight of what is required to operate an electrical appliance for useful work.

The voltage is the pressure or force, which pushes the **electrons** (-) through the resistance, or the load (s) of the circuit, and back to the voltage source (+), in order to make the appliance operate. In alternating current (AC), the path is reversed 120 times per second. Without voltage, the electrons do not move. A coulomb of electrons = one **ampere**. Amperage is the amount or quantity of electrons flowing in the circuit.

Amperage can be compared to using a garden hose to extinguish a fire. If the pressure is high, as with high voltage, then "X" amount of work can be accomplished with "Y" amperes or current. On the other hand, the same work can be accomplished with ½ the pressure or voltage, but the amperes will be 2 X "Y" or twice as much.

Example: A dual voltage, ½ Hp motor wired to operate on 120 volts, will draw 5 amperes.

The same motor, however, wired on 240 volts will draw only 2.5 amperes.

Using the mathematical relationship, Voltage times Amperes = Watts, the power consumed is the same (120 x 5 = 600, and 240 x 2.5 = 600) amount of Watts. **Watts** are what you pay for, not volts or amperes.

It is how well a device converts electricity into useful work determines its efficiency. There are many appliances, electronic and electrical devices that operate at higher efficiencies than other similar products available. Manufacturers may offer multiple products that do the same job. The consumer should compare the product's actual energy consumption data to determine which product is most efficient.

Key Words

Watt
Voltage
Electron
Ampere

Nuclear Energy

Nuclear power plants came to life in 1954. In the United Stated public rejection and government mandated safety standards, made the acquisition of permits to build a nuclear power plant was and is a long process that takes years of planning. Opposition to the use of nuclear energy, stemmed from fear of *reactor* failure causing an explosion and / or reactive fallout. Through education and the safety records of operating nuclear power plants this fear has diminished.

The main difference between a nuclear power plant and a fossil fuel plant is the method used to heat the water or to produce steam. The operations of the generators are the same. Nuclear power plant workers are exposed to less *radiation* than a person flying on an airplane or being X-rays for medical diagnostic purposes.

Using nuclear energy makes it easier for the utility companies to comply with the Clean Air Act of 1970. Pollutants and greenhouse gases produced by the coal and natural gas power plants have a greater environmental impact than spent *uranium fuel rods*. Life-cycle emissions from nuclear energy are comparable to that of electrical power produced by hydroelectric, wind and solar. Coal and natural gas produce over 600 tons of carbon dioxide per gigawatt-hour.

There are 104 operating nuclear reactors in the United States providing approximately 20 percent or 806 billion kilowatt-hours of the United States electrical requirements. These nuclear power plants are the No. 1 source of emission-free electricity. Thirty countries worldwide operate 439 nuclear reactors for electricity generation and are in the process of constructing 35 new nuclear plants. In 2007, these reactors produced approximately 15 percent of the world's electrical needs.

The used fuel rods are safely stored on site in water pools at the plants. There are plans to use new technologies to recycle the spent fuel rods recapturing the remaining energy and reducing toxic byproducts. Nuclear waste is also generated by the medical field.

With the supply of oil and natural gas dwindling, and the limitations placed on some of the other methods of producing electricity, nuclear power in place of fossil fuel power plants can provide a clean green environment.

Key Words

Nuclear power
Reactor
Radiation
Uranium
Fuel rod

Fuel Cells

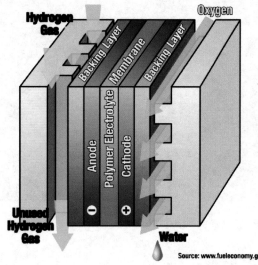

Source: www.fueleconomy.gov

Fuel cells technology has been around since 1839 and NASA is one of the largest users. NASA uses the fuel cells to provide electrical power in their spacecraft. Even though a fuel cell operates similar to a battery, it does not run down or require recharging. Fuel cells are constructed of two *electrode* plates, with an *electrolyte* packed between the two plates. As *hydrogen* (fuel) is sent to the *anode* (+) of the fuel cell and oxygen is passed to the *cathode* (-), the cell will generate electricity. The two bi-products are heat and water. This heat can be used for other heating purposes, depending on the application.

Many types of fuel cells have been developed. These cells can use hydrogen from various sources, including gasoline, methanol, methane, natural gas or any other hydrocarbon fuel resources. Many of the fuels must first be reformed or refined to produce hydrogen. Some fuel cells are fed by compressed hydrogen and oxygen. These cells are more efficient. One commonly used cell is the Polymer Electrolyte Membrane, which is a proton-exchange cell.

Hydrogen Reformation is the method used for converting the hydrocarbon materials into hydrogen.

Fuel cells can be arranged or *"stacked "* to produce the needed voltage and current requirements for large applications, including automobiles and commercial equipment.

A fuel cell will leave a much smaller carbon footprint than a fossil fuel combustion process, because it uses a chemical process to operate. For this reason fuel cell are environmentally friendly. The future energy source for transportation, electrical power, lighting, and countless other applications, may be this Green alternative.

Key Words

Electrode
Electrolyte
Hydrogen
Anode
Cathode
Stacked

Photovoltaic

The **Photovoltaic** (PV) process is now being used to power street lighting, NASA satellite equipment, calculators, watches, roadside signs, and even provide power to the electrical grid. It seems as though there is no limit to the possibilities, and best of all, the power is environmentally friendly, clean, quiet and saves enormous amounts of natural resources. The power comes directly from sunlight, through the solar cell circuit, to the user. "**Photo**" means light, (from the sun) and "**voltaic**" refers to the production of electricity. Solar **cell**s are photovoltaic devices that can be manufactured in various configurations or modules, such as those found on rooftops, at ground level and mounted solar cell modules.

Key Words

Photovoltaic
Photo
Cell
Power Grid
Semiconductor
Array

The solar cell consists of at least two layers of semiconductor materials. One of the layers has a negative charge and the other is positive. The semiconductor atoms absorb some of the photons from the light that enters the cell. This action frees the electrons from the cells' negative layer. The electrons pass through an outside circuit, back into the positive layer. Electron flow is "how electrical current is produced". The modules of PVs can be wired together in series or parallel configurations for higher voltage or higher current capacity; series for higher voltage and parallel for higher current.

Photovoltaic power can be used to charge or re-charge batteries, which in turn, are used to power electrical devices.

When modules are interconnected, they are called a Solar Photovoltaic **Array**. A high percentage of the power produced by the Solar Photovoltaic Arrays is sent to the distribution **power grid**.

These photovoltaic solar panels can produce the voltage needed for many practical uses, and present a very Green alternative.

Parking ticket meter

Wind Turbines

Wind turbines are different from windmills. Windmills covert energy from the wind into usable mechanical work. Wind turbines produce electricity. Wind is caused by the interaction of sunlight on landmasses that create high and low weather areas.

Increasing the wind speed one mile per hour, increases the amount of power derived from a wind turbine by three fold. Wind speed can exceed a turbines capacity. Brakes and changes in the pitch of the blade prevent wind turbines from turning too fast thereby controlling speed.

Some Wind farms can cover an area of several square miles or more, with many wind turbines providing electrical power to the grid. The power produced by the wind turbines must be altered for use by the grid; the frequency and power factor must be changed by capacitor banks in order to make the generated power usable by the grid. With the proper amount of wind, a single wind turbine can produce the electricity required to operate a single house or small business.

Wind turbines can produce both alternating current (AC) and direct current (DC). DC wind turbines are generally used to produce electricity for storage in batteries, whereas AC wind turbines are used for immediate applications.

The use of a wind turbine for a specific application requires a match between the amount of electrical power required and the size of the turbine. Wind turbines are sized by the kilowatts they generate. Because wind turbines only work when the wind blows, the size of the turbine is determined by using the average wind speed. To obtain the average wind speed, a wind evaluation or study needs to be done at the location.

The term Wind Farm refers to the business of generating, for sale to the grid, power via multiple wind turbines. In many instances, the same farm is used for agriculture or cattle grazing. Most of the farms are owned by traditional power companies. The farms are generally located in areas that are free of trees, structures, and obstructions.

There are single turbine wind systems available, which are designed for homeowners and small businesses. The wind turbine is mounted at the top of a tower and must have wind speeds available of at least seven to ten miles per hour. Since there will be times when the wind will be below the stated minimum rating, the home or business should have backup power. When the winds increase to above the minimum rating, the turbine will switch back proportionately from the backup source to the turbine. At times when the turbine is generating more than needed by the user, the extra power can be automatically sold to the local utility provider.

The wind turbine is an example of using kinetic energy from the wind to help resolve some of our massive energy problems and still keep our environment Green.

To reduce losses caused by interference between turbines, a wind farm requires roughly 0.1 square kilometers of unobstructed land per megawatt. A 200 MW wind farm might extend over an area of approximately 20 kilometers or 7.7 square miles.

Key Words

Wind Turbine
Windmill
Wind farm
MW
Capacitor
Kinetic energy

Motor Efficiency

Motors used in our everyday lives consume large quantities of electrical energy. Electric motors are used to move air, pump fluids operate machinery, etc. Mechanical work requirements within a building are never constant. Controlling motor speeds provides a solution to the changing needs of the structure. In the past, controlling motor speeds on a large scale was only applied to large commercial systems, now with rising energy costs, and the advancement of electronic controls, speed control technology is being used in small residential and commercial applications.

Having an understanding of the relationship between the energy used, and the power needed to maintain work needed, motors and controls can be used to reduce energy consumption. As energy requirements decrease, the power consumption of the motor will also decrease and result in significant energy savings. *ECM*, VFD and VSD controlled motors are available for energy saving alternatives.

Electronic Commutated Motor (ECM)

Compared to other single-phase motors, ECM motors have the highest efficiency rating and are used in many residential and commercial applications. The ECM motor is a Direct Current (DC) motor, but it has no brushes. It is controlled by a built-in variable frequency drive. The frequency output of the drive provides the variable speeds of the motor, which can be adjusted for the demand, as required. When the EMC motor is used within an air handler, the dip switches on the motor's control board can be set to meet the cooling or heating *cubic feet per minute* (CFM) requirements. When used with the newer electronic thermostats (which monitor the indoor relative humidity as well as the temperature) the blower speed will vary to help control the humidity. The CFM of air can also be varied if there are multiple stages of heating and cooling. The motor also saves energy by running at 50% of the cooling speed when the thermostat fan switch is set to the continuous position.

Variable Frequency Drives (*VFDs*)

Variable *Frequency* Drives are typically found in most high efficiency commercial systems today. VFDs operate motors more efficiently by increasing or decreasing the speed of the motor. This is accomplished by changing the voltage and frequency.

Variable Speed Drives (*VSD*)

Variable speed drives can be used on DC (***direct current***), or with single phase and 3 phase AC (**alternating current**) motors. They differ from variable frequency drives in its switching action. The principle mode of operation is pulse width modulation. Pulse width modulation is the amount of time the switches are opened and closed to vary the motor speed. The controls can range in voltages from 110v to 10Kv.

Key Words

ECM
VFD
ECM
Frequency
Direct Current
Alternating current

Lighting

Low energy consuming, high efficiency *Fluorescent* and LED lighting sources are readily available due to improvements in both technology and materials.

Historically, lighting has been purchased based on the wattage. A 100-watt *incandescent* bulb produces a great deal more *lumens* (light) than a 60-watt incandescent bulb.

Modern bulbs are rated in lumens. One lumen is the amount of light on one square foot of surface created by a candle at a distance of one foot. A 60-watt incandescent bulb produces approximately 850 lumens of light. When comparing different types of bulbs, the higher the lumens the greater the light output.

Fluorescent lighting

Fluorescent lights work by using high voltage and high frequency electricity to create *ultraviolet* light that reacts with the *phosphor* coating inside the tube. The Fluorescent coating can be made of a mixture of many different materials to produce various colors of light.

Fluorescent Tube

High voltage transformers called *ballasts* are used to illuminate standard fluorescent tubes. Modern ballasts do not use wound transformers. Instead, the ballasts are made with electronic components that reduce the amount of electricity consumed. A standard dimmer switch cannot be used to control fluorescent lighting. Instead, an electronic *dimming ballast* must be used to allow the fluorescent tubes to operate at lower power levels. Thus reducing the amount of light when total illumination is not necessary will save energy.

Compact Fluorescent

Compact fluorescent lights are designed to fit in the same socket as a standard incandescent bulb, allowing most incandescent bulbs to be replaced without having to replace the light fixture. The tubes of compact fluorescent lights are bent and curled to occupy the same space as a standard incandescent bulb. These compacts operate in the same manner as fluorescent tube type lighting. The electronic ballast is located in and above the screw threads, before the bulb.

Fluorescent lights use about 1/4 of the electrical energy when compared to a standard incandescent light with the same lumens output. The life expectancy is considerably greater than standard incandescent bulbs. Over time, fewer bulbs are replaced and the amount of electricity consumed is reduced.

Example;

An incandescent bulb uses 60 watts of power to produce approximately 850 lumens of light, and has a life expectancy of 1500 to 2000 hours. This bulb will produce about 14 lumens of light per watt of power. A compact fluorescent light uses 13 watts of power to produce approximately 900 lumens of light, and has a life expectancy of up to 10,000 hours. This bulb will produce about 69 lumens of light per watt of power. When comparing lighting, always compare the number of lumens to the actual wattage.

Key Words

Fluorescent
Ultraviolet
Phosphor
Ballasts
Incandescent
Lumen
Compact
    Florescent
Dimming ballast

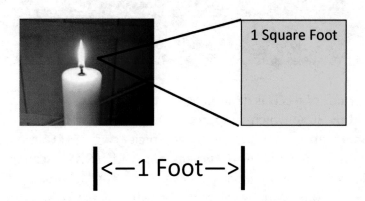

1 Square Foot

|<—1 Foot—>|

LED (Light Emitting Diode)

Light emitting diodes (*LED*) have been around since the 1960s. At first, they were used as indicating lights on control panels. The color of the emitted light depends on the composition and condition of the *semiconductor* used. In 1993, a blue indium gallium chip was created with a phosphor coating which create a white light from a single *diode*. To form a full-color pixel, a cluster of red, green, and blue diodes must be used. With today's technology, grouping and changing the color patterns of LEDs are an easy task.

Due to their long life and low power consumption, LEDs are being used in new applications, such as automobile taillights, airport runway lights, flashlights, traffic lights, and billboards.

Compared to fluorescent and incandescent bulbs, LED lighting is less prone to damage from thermal and vibration shocks. These lights do not produce heat or flicker and perform well when subjected to frequent on-off cycling. Unfortunately, LEDs are very sensitive to excessive heat, and when exposed to inappropriate applications the LED will dramatically reduce in both light output and life span.

The "useful life" of a LED is defined as the time it takes for the initial light output to be reduced by 30%, which is approximately 50,000 hours. An incandescent bulb will last approximately 2500 hours. The efficiency of a LED in a typical residential application is approximately 20 *lumens* per watt (*LPW*). Incandescent bulbs have an efficiency of approximately 15 LPW and compact fluorescents are approximately 60 LPW, depending on the wattage and lamp type. LEDs are better at placing light in a single direction than incandescent or fluorescent bulbs. There are LEDs for commercial use that approach 75LPW.

Bulbs for low lumen or wattage applications have been developed. These include standard bulbs, flood lamps, spotlights, candelabra and candle-type bulbs. There are flashlights which use LED's , and can be used continuously for as long as 150 hours before fresh batteries are needed.

With the world seeking newer ways to reduce carbon emissions, lower energy consumption, and going "Green", replacing just a few lamps in a building can save on the energy required for lighting and also reduce the heat load for the air conditioning system.

Key Words

LED
Diode
Semiconductor
Lumens
LPW

Tidal and Ocean Energy

Key Words

Tides
Tidal turbine
Current turbine

Electrical power generated from ocean waves, *tides*, ocean currents, and even river currents are renewable and sustainable Green Alternatives. Other alternatives include off shore windmills and ocean thermal gradient energy resources. These exciting, new projects have resulted in vast amounts of technical data is being gathered, such as how much electrical energy can be generated and the problems that will be encountered by time and nature, as well as solutions. Each of these resources has their own unique assets and potential.

Turbines use energy of the motion of a fluid, such as air or water, and convert that energy into electricity. The denser the fluid is, the greater the power output per rotation.

Waves and tides

As taught in nearly all grade schools, the ocean has vast amounts of power stored in the never-ending tides and waves, created by nature. The tide changes sea level due to the Earth's rotation. The effect produced by the gravitational forces of the Sun and Moon. The tidal effect diminishes as you approach the equator. Much of this energy can be converted into mechanical and electrical energy. Unlike the wind, which is necessary for wind turbines, tides are both predictable and reliable. Power generation occurs when the tide flows and ebbs. The magnitude of energy generated depends on the intensity of the tides and waves.

Rivers

The fast moving waters in rivers are another resource used to produce energy for electrical power. The river's current propels these submerged turbines. They are not, however, completely free of potential problems in this riverbed environment. Rivers currents often carry a variety of debris and sediment, which can produce potential problems for *current turbines*.

Energy Extraction Devices

Just as there are various windmill energy extraction devices, there are also many different designs for the ocean and river turbines. Using the ground as a reference, some are horizontally mounted and others are vertically mounted. Most of these systems have a rotating blade, which is turned by the energy of moving water. *Tidal turbines* work well when the tidal current is at least five knots. The tide has four six hours nearly continuous intervals in each direction and is very predictable. This makes tidal turbines more adaptable for connecting to an electrical power grid than wind turbines, which turn only when sufficient wind is present.

Ghost Loads

Ghost loads also known as **lazy loads**, **phantom loads** or **stand-by loads** are small drains of electricity that can add dollars to your electric bill. These power consuming loads include instant on circuits in televisions, anything that uses a transformer such as a cell phone charger that is plugged in but not in use, etc. These devices draw and consume power as long as they are connected to a power source. The use of a simple switch between the power source and the device can result in a reduction in your electric bill.

Key Words

Ghost loads
Lazy loads
Phantom loads
Stand-by loads

Residential Major Appliances

The term "residential appliances" refers to every electrical item used by and in the household, from the central air conditioning and heating system to the electric toaster and lawn mower. Appliance manufacturers often provide energy savings tips within their product literature. Many appliances can be simply turned off when not in use; others should be disconnected from their power source to eliminate "Ghost Loads".

Major appliances, such as washers, clothes driers, and water heaters can be the real energy culprits. Before a new major appliance is purchased, the buyer should review the yellow Energy guide label attached to many of these appliances. This tag shows the key features of the appliance, the estimated yearly operating cost, and cost range of similar models. It also has the kWh or estimated yearly electricity use.

Key Words

humidifiers,
dehumidifiers

Estimates of operating costs for nearly all appliances can be found at numerous web sites or the following formula can be used to calculate the cost, if the cost per kilowatt hour from the utility is known. The yellow tag is based upon a national average electricity cost of 10.65 cents per kWh.

For example, an appliance having 1500 Watts of power, on continuously for 5 hours, would use 7500 Watt hours of energy.

This is 7.5 kilowatt hours
7.5 kWh, multiplied by the rate of 10.65 cents equals a cost of $ 0.80.

Watts÷1000 x hours = Kilowatt-hours (kWh) x rate = cost
1500 watts÷1000 x 5 hours = 7.5 kWh x 0.1065 = .79875 or $ 0.80

Dishwashers, refrigerators, freezers, clothes washers, water heaters, HVAC equipment, and pool heaters have the Energy guide label attached.

Some appliances, however, *do not* have the yellow tag Energy guide label on them.

These are:
Electric ranges, televisions, ovens, humidifiers, dehumidifiers, and Clothes Dryers.

Green Plumbing Systems

Key Words

Hydrologic
Potable
Hydrologic cycle
Hydrologic system

The design of a building's plumbing or ***hydrologic system*** is an integral part of constructing a sustainable or green building. Green plumbing systems not only conserve the precious resource of water, but also reduce the energy requirements for moving, treating and heating that water.

Moving ***potable*** water from the source to the point at which it is used requires energy. Large cities often pump water great distances to meet the needs of their inhabitants. The greater the distance potable water is moved the higher the energy demand. Electrical power plants usually provide this energy. Many of these power plants burn fossil fuels to create electricity. Using less water for plumbing or irrigation purposes lessens the amount of wastewater needing treatment and the energy needed to accomplish this treatment, thereby decreasing the carbon footprint.

A building's ***hydrologic cycle*** begins with the use of potable water for drinking, washing, cleaning and waste transport. After use the previously potable water, now wastewater is transported to a gray water or reclaimed water reuse system or to a public or private sanitary treatment system. After treatment, the effluent is discharged into disposal fields or surface water sources such as oceans, rivers, and lakes. Water treatment facilities take water from aquifers and those same rivers and lakes and produce potable water, which is distributed to homes and buildings creating an endless cycle of use and reuse.

The hydrologic cycle also includes the building's systems for landscape irrigation and storm water removal or use. Plumbing systems involved in this cycle are the building water supply distribution piping, irrigation piping, fire sprinkler piping, drainage piping, and storm water piping. Each of these piping systems should be addressed in a true green plumbing system.

Water conservation is the goal of green plumbing design. Water resources in many parts of America are taken for granted. As population increases areas, where water is plentiful will see a strain on their water resources if conservation measures are not implemented. Of all the water on our planet less than 2.5 percent is fresh water three quarters of which is frozen or otherwise not attainable. Only 0.3 percent is available for our use in the form of rivers and lakes and most of that resource is polluted requiring treatment before drinking. By reducing the use of potable water in buildings, which currently accounts for 12 percent of the use of fresh water, we will greatly enhance the sustainability of our drinking water supply by saving billions of gallons of water each year.

In many parts of the world, drinking water is a precious commodity not to be squandered. The less water we contaminate with pollution and the more water we conserve can help provide assistance to the 1.1 billion of the world's population without fresh drinking water and the 2.2 billion without sanitary services.

Reducing the amount of potable water used through innovation aimed at conservation thus has become more and more critical. It can also be one of the simplest and most economical methods of conservation. A homeowner at a reasonable cost can accomplish most of the methods for reducing water use described below. They most certainly should be on the top of a service plumber's list for recommendations to a client and they soon will be, if not already, common practice for new construction projects. At the very least leaking faucets, fixtures and irrigation systems must be repaired saving thousands of gallons of water per year.

Adapting an existing home or building drainage system to one of the following *wastewater* reuse systems can be difficult and costly; however, it can still be accomplished. Small packaged retrofit systems are being designed and sold. The installation of this kind of gray water or reclaimed water system should be left to a qualified plumber. Particular attention to installation and strict adherence to plumbing codes is necessary. Most green building construction will likely include some form of wastewater reuse system.

Storm water control and *rainwater* harvesting were among the first methods used to protect the environment. Storm water drainage systems, were designed to protect villages and towns from destructive flooding. It was only later that the convenience of transporting human waste came into play eventually leading to the development of fixtures and plumbing drainage systems. This ultimately led to the pollution and depletion of our water resources to the point where we find ourselves today attempting to conserve water by rainwater harvesting – one of the original methods for providing water for drinking water and other uses.

In addition to rainwater harvesting, storm water (rainwater falling on surfaces other than a roof) control, catchment, and treatment has again become an important part of protecting the environment. Storm water washing over parking lots or roads can carry up to 4 gallons of oil and gas per acre per year. Collection and treatment of this polluted storm water is critical for the protection of rivers, lakes, and oceans.

Fire sprinkler systems can also be included in green plumbing systems. The fire suppression system reduces the affects of fire damage and its resultant debris. Mist systems can replace deluge systems conserving water and limiting water damage when extinguishing a fire. Halon fire suppression systems are also being replaced to reduce the impact of Halon on ozone.

Including green plumbing systems in a sustainable building design and increasing water, efficiency provides economic, societal, and environmental benefits. It can lower initial installation costs and reduce annual energy, water, and sewage costs. It preserves our precious water resources for future generations and for use for agriculture and recreation while reducing the need to expand or build new potable water and wastewater treatment plants. Environmentally green plumbing systems lower potable water use and the wastewater discharge thereby decreasing the strain on aquatic ecosystems and preserving water resources for wildlife and agriculture.

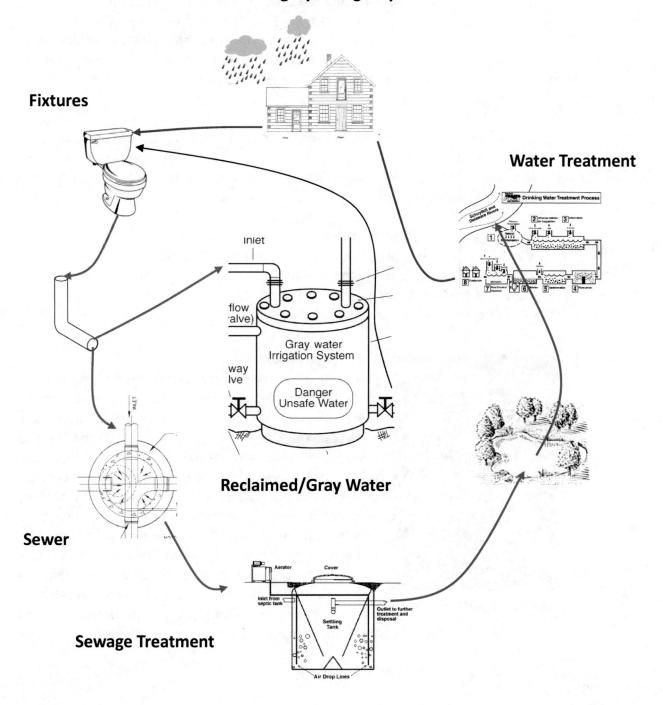

The Building Hydrologic Cycle

Fixtures

Water Treatment

Reclaimed/Gray Water

Sewer

Sewage Treatment

Potable Water Conservation

Potable water systems are plumbing systems that supply water suitable for drinking to plumbing fixtures, dishwashing, cleaning, laundry, and waste transport systems. In 1992, the Energy Policy Act was passed requiring that all plumbing fixtures and fixture fittings used in the United States meet targets for reducing water consumption. Water closets 1.6 gal per flush, urinals 1 gal per flush, showerheads 2.5 Gpm @ 80 psi, faucets 2.5 Gpm at 80 psi, replacement aerators 2.5 Gpm and metering faucets 0.25 gallons per cycle. Since that time, plumbing codes have mandated these requirements. The goal of a green plumbing system is to reduce the water flow below the EPA 1992 amounts.

Key Words

Aerator
Potable
Energy Policy Act
Water closet (WC)
Urinal
Faucet
Showerhead

Conserving water by creating a more efficient supply can be the most effective and economical method of greening a building. Reducing the use of water and meeting the goal of green plumbing can be achieved in any residence or building by the following methods:

- Outlet or faucet flow restriction
- The installation of high efficiency plumbing fixtures and fixture fittings
- Replacing wasteful plumbing appliances
- Improving the efficiency of the hot water distribution system
- Improving the landscape irrigation system

Aerator with Valve

Flow restriction

Flow restriction devices can be attached to many types of fixtures and fixture fittings. One of the easiest and most effective are replacement faucet aerators and shower heads. These devices reduce the volume of water being used. Many of these devices are simple restriction devices or orifice plates. Water flow is restricted because of the size of the hole in the plate. Advanced technology aerator devices now have pressure compensation features that maintain desired flow rates when pressure is decreased. They mix air with the water, thereby increasing velocity, which gives the impression that the flow has not changed. Care should be taken in selecting the flow restriction device so that the performance of the fixture or faucet is not adversely affected.

Faucets

Lavatory and kitchen faucets can use as much as 15% of the water used in a home. Low flow aerators can be attached to most of these lavatory and kitchen faucets. The aerator mixes air with water to give the appearance that there is a large volume of water. Every aerator has a flow rate stamped on the side. Models can be found that limit the flow to less than one Gpm. Another type of aerator incorporates a valve into the device that can stop the flow of water completely conserving even more water. The valve should not be utilized until after the arrival of hot water or this could increase the wasting of water by waiting for hot water to arrive while the cold water goes down the drain.

Showerhead

Showers can use as much as 22% of the water used in a home. There are many designs of low flow showerheads. Some of these maintain a high-pressure spray while others try to imitate rain. Spray types generally include concentrated, pulse, rotating, and gentle sprays. Consumers have a number of choices to meet their personal preference. What is important is the amount of water flow per minute. A high efficiency showerhead uses approximately 1.5 – 2.3 gallons of water (8 - 9 liters) per minute or less. Some models incorporate valves designed to stop the flow of water completely. Hand spray showers also have this feature. Care should be taken to ensure that the new showerhead does not adversely affect the anti-scald features of the shower valve.

Low Flow Showerhead

Many plumbing manufactures produce shower systems with multiple showerheads. These systems can easily double or triple the amount of water used. Plumbing Code development committees and several localities around the country are in the process of placing limitations on multiple showerhead systems.

Pre- rinse spray valves

Pre- rinse spray valves are used in commercial kitchens to rinse food containers before washing. Prior to 2006, the average spray valve used 3-3.5 gallons of water per minute. In 2005, the updated Energy Policy Act included a maximum flow rate of 1.6gpm for pre-rinse spray valves. There are models now available that use less than 1.3gpm. Using a low flow spray valve could save up to 180 gallons or more of hot water over a three-hour use period. The pre-rinse spray valve should be installed with an integral shut off valve, which will provide further water savings. The reduction in energy and water costs can recoup the cost of the valve in a few weeks.

Reducing water consumption and the amount of energy needed to pump, condition and heat water is the objective of low flow devices. A typical 10-minute shower uses 42 gallons (190 Liters) of water. High efficiency showerheads can dramatically reduce the amount of water and energy used. Other high efficiency water devices reduce the impact on water resources and save energy.

Low Flow Pre-Rinse Sprayer

High Efficiency Plumbing Fixtures

Key Words

Ultra Low Flush (ULF)
WaterSense
Flushometer tank
Flushometer
Dual Flush

High efficiency plumbing fixtures push the limit of low water usage to no water usage. Reducing the amount of water used for wastewater and fecal matter transport has the potential to save hundreds of gallons of fresh water per day.

Aerators, showerheads, and other flow restricting devices are used to provide reductions in water use for sinks, lavatories, bathtubs, and showers. For water closets and urinals, the fixture itself must be redesigned for lower water usage.

Just reducing the amount of water used per flush will not in itself reduce water usage. A water closet that does not completely remove the waste and clean the bowl will be flushed repeatedly, increasing in the amount of water used. Therefore, the object of these low flush fixtures is to completely flush and wash the bowl with as little water as possible.

Water closets (WC) and urinals labeled as high efficiency or *ultra low flush (ULF)* reduce the amount of water per flush below the 1992 Energy Policy Act requirements of 1.6 gallon per flush for a water closet and 1.0 gallon per flush for a urinal. To be sure that a ULF WC or urinal is designed properly make sure that it is *WaterSense* labeled. WaterSense is a partnership program sponsored by the U.S. Environmental Protection Agency to promote water-efficient products and practices across the country.

Water closets (toilets)

Water closets account for approximately 27% of the water used in a home and almost 50% of the usage in a commercial building, not including irrigation. Almost 4.8 billion gallons of water per day are flushed down the drain through water closets. Many WCs in use today do not meet the standard of 1.6 gallons per flush set by the 1992 Energy policy Act. Many older buildings and homes are equipped with 3.0 gallon or higher per flush water closets. Replacing these older WCs with 1.6 Gpf WCs could save almost 5,500 gallons per person per year. Replacing older WCs with new high efficiency Ultra Low Flush and Dual Flush fixtures will save thousands of gallons per day.

Flush Tank WC

It is important to note that the installation of these high efficiency water closets is virtually the same as the 1.6 Gpf or higher WC. Some of these WCs have a larger trapway of 2-1/2 to 3 inches but the connections are identical to other WCs. The water supply line is the same size as is the fixture's connection to the closet flange.

Ultra Low Flush Water Closets

Ultra Low Flush (ULF) WCs reduce the amount of water used from 1.6gpf to 1.28gpf or less providing a usage decrease of almost 20%. This could offer a typical family a savings of 4,000 gallons of water per year.

These water closets are flush tank, *flushometer tank* (pressure assist) or *flushometer* type water closets. Each uses a siphoning action to empty and wash the bowl of the WC. ULF or high efficiency water closets are available for residential and commercial applications.

Flush tank or gravity flush type WC's, utilize a flush valve at the base of the tank to release the flow from the tank to create a siphon in the WC trapway evacuating the waste while washing the bowl. With this type of WC, the waste is pulled or siphoned out behind the flow of water.

The flushometer tank WC or pressure assist type uses water to build pressure against an air bladder inside the tank. When flushed, air in the bladder expands into the trapway and rim water ports providing extra force pushing the tank's water into the bowl and its contents out of the bowl. The waste in these WCs flows in front of the flush water providing for somewhat better transport of solid waste.

The flushometer WC is used primarily in commercial applications. It utilizes a pressurized flow directly from the water distribution system through a flushometer valve, to create a siphon action to empty the waste and clean the bowl. Some of these WC's utilize a blow out bowl where the flow of water pushes out into the bowl pushing the waste through the trapway. Flushometer valves meter the proper amount of water for each flush. The addition of an electronic sensor adds a convenient touch less activation to the flushometer valve.

Flushometer Tank WC

Flushometer WC

Key Words

Dual Flush

Dual flush water closet

A *Dual flush* water closet provides the user with flushing options. The dual flush WC is equipped with two buttons on top of the tank lid or a two-position handle on the side of the tank, which allows the user to choose which volume of flush to use. One option initiates a 0.8 to 1.0 gallon flush for liquid waste only. Water is saved by emptying only the contents of the bowl without having to create a large siphon to empty solids. When solids are deposited in the bowl the second option allows for a 1.28 to 1.6 gallon flush to remove the solids and clean the bowl. The reduction of water use can be significant with up to an additional 67% of water saved using dual flush technology

Dual Flush WC

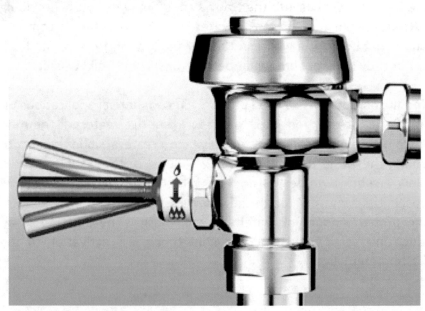

Dual Flush Flushometer Valve

Ultra Low Flush Urinals

According to the 1992, EPA the urinal flush requirement is 1.0 gallon per flush. *Ultra low flush urinals* reduce the flow per flush to 0.5 gallons or less. The object is to allow water to wash down the urinal and provide transport of the highly acidic urine. These urinals can provide a 50% or more reduction in the use of water.

Key Words

Ultra low flush urinals

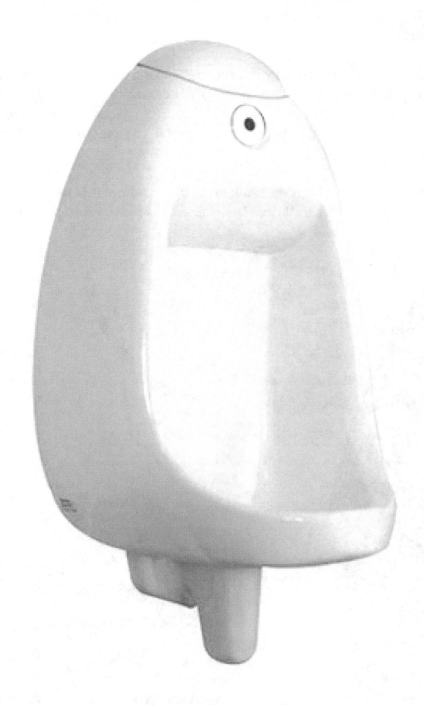

Low Flush Urinal

Waterless fixtures

Waterless fixtures were the original fixtures for the disposal of human wastes. Chamber pots were used for centuries to contain and then remove waste from buildings. The waste usually was disposed of by throwing it out the window or door onto the street. Towards the middle of the 19th century, fixtures began to be placed inside homes and piped to privy boxes in basements, which were then emptied. The waste was deposited in the most convenient places usually lakes, rivers or streams.

Waterless fixtures are used extensively around the world and have been proven to work effectively. Waterless, no-flush urinals, and *composting toilets* do not use water to flush away human waste. This can lead to significant water savings but also requires significant maintenance and careful consideration of plumbing codes.

The waterless urinal is designed to allow urine to pass through a trap or trap like device without using water to flush. The trap area is designed to allow urine to

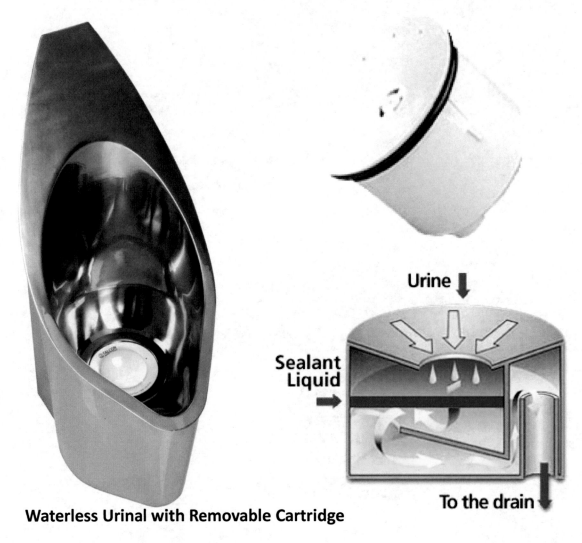

Waterless Urinal with Removable Cartridge

pass through to the drain but seals out sewer gas. In many cases, the trap or device can be removed from the urinal for maintenance. Regular maintenance requires a light spraying of cleaner and wiping down of the urinal each day with periodic replacement of liquid sealant. The waterless urinal is installed in the same manner as a water supplied urinal including the drain piping and size.

Removable Cartridge / Insert Waterless Urinal

Removable *cartridge* or *insert* type waterless urinals, as other urinals, are made from china, fiberglass, or stainless steel. A replaceable cartridge or insert is installed at the bottom of the urinal into a fitting that is connected to the fixture drain. The cartridge features a sealing ring to create an airtight barrier and allow easy replacement. The cartridge works as a funnel ensuring that the urine passes into the cartridge and through a liquid sealant that floats on residual urine below it. The liquid sealant provides the trap seal preventing sewer gas from penetrating the cartridge and entering the room. Particular care must be taken to ensure that the sealant is not removed and that the cartridge in maintained properly. Proper cleaning and sanitation of the urinal is essential for odor free operation.

Cartridge Free Waterless Urinals

A *cartridge free waterless urinal* has an integral trapway and trap that slows the flow of urine through a trap that contains a liquid sealant. This ensures that the flow does not wash the liquid sealant out of the trap. The liquid sealant is used in the same manner as the cartridge type urinals. The sealant floats above the residual urine in the trap maintaining a barrier to block trapway odors. The fixture also has an integral screen to keep solids from entering the trap. Proper maintenance once again is critical. The liquid sealant must be replenished on manufacturer specific intervals along with periodic flushing of the trapway with water. The urinal must also be wiped down on a periodic basis.

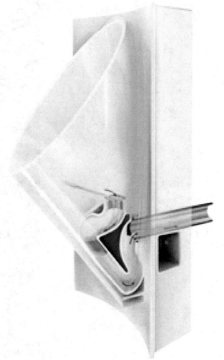

Cartridge Free Waterless Urinal

Waterless urinals are not free from controversy the fact they allow urine to flow through a drainage system without dilution or water transport until meeting a drain with water flow. There has been evidence of solids buildup in the drain lines from waterless urinal's causing increased drain line maintenance. The lack of water to wash down the fixture itself leads to reliance on proper maintenance intervals to keep the urinal clean and odor free. Due to these facts, many installations of the waterless urinal have been replaced by water-supplied urinals. Consequently some plumbing codes require water supply piping be installed behind the wall supporting the waterless urinal for possible water supplied urinal replacement.

Key Words

─────────

Composting toilet

Composting Toilet

The ***composting toilet*** is very different from the water closet and must not be confused with an outhouse toilet. The composting toilet consists of either a combined seat with a self-contained composting compartment or a toilet with a remote composting tank, which provides the area for composting of fecal matter. The composting toilet converts the deposited human waste into an organic compost and usable soil through the natural breakdown of organic matter into its essential minerals without the use of water. Aerobic microbes accomplish this in the presence of moisture and air, by oxidizing the carbon in the organic material to carbon dioxide gas, and converting hydrogen atoms to water vapor. Composting eventually reduces the volume of waste to approximately 1-2% of the original volume of waste.

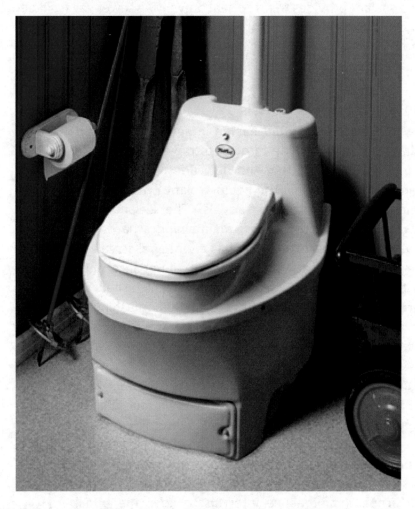

Self Contained Composting Toilet

Composting toilets range from simple twin chamber designs to advanced systems with rotating tynes, temperature probes, moisture probes and electronic control systems. They are effective biological converters of human and household "waste," saving money and energy for the person and community.

The composting toilet does require some removal of compost product although a properly designed and installed system will require this on only a biannual or quarterly basis.

The installation of this toilet will be quite different from the normal water closet. Requiring strict adherence to manufacturer's instructions, compliance to local codes and health department regulations will also be critical. Many codes do not have provisions for these toilets. Be sure, to consult the local authority having jurisdiction over composting toilets for approval prior to installation.

This toilet may only be desirable in unique installations such as national park facilities. Composting toilets are used for example in the Grand Canyon National Park.

Each waterless urinal can save as much as 45,000 gallons of water annually. The composting toilet can eliminate the use of water saving thousands of gallons. This staggering amount of fresh (potable) water savings can result in a tremendous energy reduction.

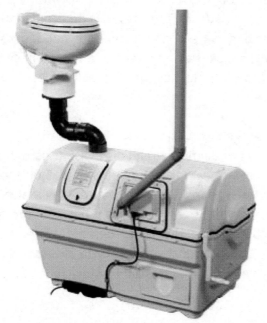

Composting Toilet with Remote Tank

Key Words
_____

Clothes washer
Energy Star
Dishwasher

High Efficiency Plumbing Appliances

High efficiency plumbing appliances refers to **clothes washers**, **dishwashers**, and commercial ice machines. The **Energy Star** Label can easily identify a high efficiency plumbing appliance. An Energy Star model appliance incorporates advanced technologies that use 10 - 50% less energy and water than other models.

Clothes washers use approximately 22% of indoor residential water. Conventional washing machines can use up to 40 gallons of water per load. A high efficiency or **Energy star** model can reduce the water used in half. A washer that meets the Energy Star model criteria can come in either front loading or top loading models.

Front Loading clothes washers tumble clothes through a small amount of water instead of rubbing clothes against and agitator in a full tub. Advanced top loading clothes washers flip or spin clothes through a reduced stream of water by a sophisticated new system. Both designs use less hot and cold water in the wash cycle thus saving water and the energy needed to heat it. Efficient motors spin clothes much faster during spin cycles to extract more water from the clothes, which results in less drying time.

The average dishwasher uses 6 gallons of hot water per cycle. The Energy Star qualified model dishwasher uses 4 gallons of hot water per cycle and can save an average of 5,000 gallons of water per year over washing dishes by hand. It will also use approximately 41% less energy than other models. Water and energy can be saved by utilizing the pre programmed reduced time and water cycles. Using the dishwasher for full loads will also save water and energy versus washing multiple small loads of dishes.

Air-cooled **cube type ice machines** that are Energy Star qualified, are typically 15% more energy efficient and 10% more water efficient than standard models. They include self contained and remote condensing units. Energy savings can be up to 1160 kWh and 2700 gallons of water per year.

Using other Energy Star compliant **commercial appliances** can reduce energy costs. Hot food holding cabinets, commercial fryers, commercial refrigerators, and freezers as well as, commercial steam cookers are appliances that are manufactured in Energy Star qualified models. They can save anywhere from 10 to 30% in energy usage and costs per year.

Energy efficient **garbage disposals** can also be found, although, if one is looking to be truly green then the use of a garbage disposal cannot be recommended. Not only does use of the garbage disposal waste water and energy but also the addition of solid waste in the sewer system increases the amount of load and treatment needed at the sewage treatment plant. A better option is to place the garbage in a garbage can or to compost this organic material.

Hot Water Distribution System

Reductions in water usage and the energy it takes to heat water can be made by choosing the most efficient means to heat water for a particular purpose or installation. The methods and materials used to deliver hot water to individual fixtures and appliances can make a difference in the amount of water used and the energy it takes to heat that water. The entire water distribution system should be taken into consideration when determining which method of water heating should be used and how it will be piped throughout a building.

A well thought out and integrated hot water distribution system will provide the highest efficiencies in a new building. Changes to an existing hot water distribution system can be also made. Replacement of inefficient water heating equipment will make a significant reduction in water and energy usage. In most cases, it would take an entire remodel of a home or building to change the method of piping to fixtures and appliances, however, methods to *circulate* hot water can be made to reduce the amount of water and energy used in an existing installation. Simply *insulating* an existing water heater and exposed hot water piping can make a big difference.

Key Words

Circulate

Insulating

Items to consider in designing or remodeling a hot water distribution system should include:

- Insulation of water heating equipment and piping

- Hot water circulation systems

- Sizing and installation of hot water piping

- Higher efficiency water heating equipment

One of the easiest ways to make an immediate impact on energy savings is to insulate a new or existing water heater. Unless the water heater itself has an R factor of at least 24, adding insulation to the tank can reduce standby heat losses by 25-45% and save up to 10% in energy costs. Insulation packages for many models of water heaters can be found in pre-cut jackets or blankets.

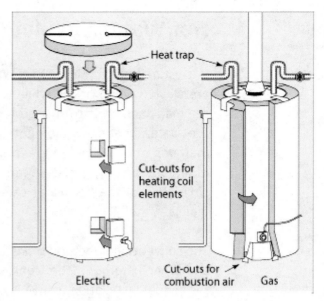

Water Heater Insulation

Hot water piping should be insulated throughout its entirety. Pipe insulation will prevent heat loss and provide sound isolation from noisy piping. It has been estimated that 1" of insulation over a 2" hot water pipe could save approximately 2.6 therms and $2.00 in energy costs annually per linear foot of piping. Insulating all of the hot water piping can be accomplished easily in new construction projects but can be near impossible in existing buildings. At the very least exposed hot water piping at fixtures, in attics and basements should be insulated. Cold water piping should also be insulated a minimum of 3 to 6 feet from the from the water heating equipment.

Pipe Insulation

Key Words

Piping method
Pipe sizing
Circulation

Hot Water Circulation Systems

It is estimated, that in a typical American home, 10,000 gallons of water per year are wasted down drains while waiting for hot water to arrive at a fixture. There are many solutions to this problem. Some of them involve proper placement of the water heating equipment in the building as well as using specific *piping method*s and popper *pipe sizing*. Changing piping methods and sizing cannot easily be accomplished in an existing building.   One solution is providing *circulation* of the hot water can be done without tearing walls out and replacing piping.

The conventional method of installing water piping in a building or home is a trunk and branch method. The trunk, larger lines normally ¾ to 1" in size in an average residential installation, conveys cold water to and hot water out of water heating equipment.   Branches, smaller lines normally ½" in size, deliver hot water to fixtures and appliances. At the fixture, this branch size is usually reduced to 3/8" in size by the fixture supply, usually flexible tubing connecting the branch outlet valve to the fixture or appliance itself. Hot water stays in the line until the fixture or appliance is used. Even if insulated, the hot water in the branch and fixture supply piping will eventually lose heat and become cool. When the fixture or appliance is finally used, the occupant normally waits until the water is hot once again until utilizing the hot water. This is indeed wasteful and unnecessary unless a method of circulating the hot water back to the water heating equipment is provided and thus eliminating the wait for hot water.

There are three methods of providing circulation of water back to the water heating equipment—

1. on demand water circulation systems
2. dedicated line water circulation systems
3. gravity circulation systems

Of the three, the on demand water circulation system is the easiest to accomplish in an existing building.

On Demand Water Circulation System

The on demand water circulation system uses a small *pump* connected between the hot and cold fixture supply, normally at a lavatory, to draw water from the water heating equipment and push it through the cold fixture supply and back to the equipment. Hot water from the heater is thus delivered to the fixture without having to drain the cool water out of the line.

Pump
Thermostatic valve
Motion detector

Some of these systems are very sophisticated including *thermostatic valves* that will turn off the pump when the water in the fixture supply reaches the proper temperature. The pump can be set to a timer or activated by a push button when needed. A *motion detector* can also be used to sense the presence of the occupant and turn on the system providing an automatic and convenient method to save those thousands of gallons of water.

The on demand water circulation system works very well for an individual fixture such as a kitchen sink. It may even work well for a group of fixtures set closely together as in a bathroom for example. However, homes or buildings with two or more bathrooms spread out apart from each other may need more than one unit to provide hot water without waste throughout the building. Installing more than one on demand water circulation system may be more cost effective than replacing piping and repairing walls.

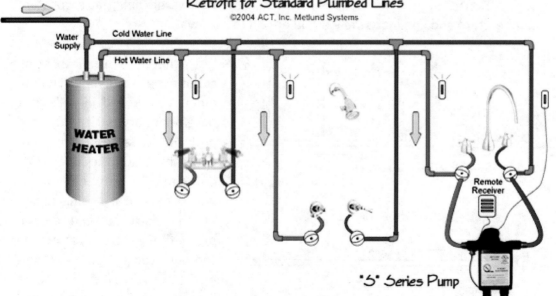

On Demand Water Circulation System

Dedicated
Timer
Thermostat

Dedicated Line Water Circulation Systems

The dedicated line water circulation system differs from the on demand system in that the circulation of hot water occurs in the piping system itself rather than at a fixture. The hot water distribution piping is designed with a *dedicated* return line from the furthest fixture or group of fixtures from the water heating equipment, which is then piped back to the equipment. Other lines from fixtures or groups of fixtures may also tie into this return line to provide hot water circulation for those areas.

The dedicated circulation line is connected to the water heating equipment at either the cold water inlet or a circulation inlet if provided. This allows the water to reenter the equipment and reheat the water in the piping. A pump is usually installed on the hot water outlet piping to push the hot water through the piping and back to the water heating equipment creating a constant circulation or re-circulating loop.

The circulation pump in the dedicated line water circulation system can be thermostatically controlled, placed on a timer or left constantly running. The pump that is left constantly running will waste energy while running when hot water is not needed. However, in some large commercial installations it may be desirable to leave the system running if hot water use occurs at all times. In most instances, the pump can be controlled by a *timer, thermostat* or both for greater efficiency. This will allow the pump to be used only during the time of day when it is needed and when the temperature in the return piping is below the thermostat set point. This system will keep the water in the hot water piping at a constant temperature and eliminate the waiting period for hot water.

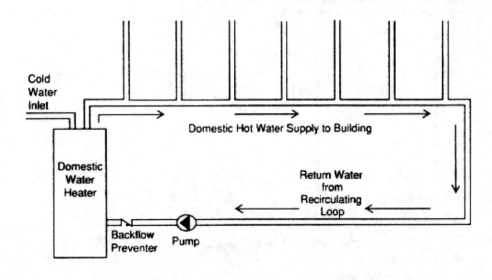

Dedicated Line Water Circulation System

Retrofitting an existing commercial building where piping is exposed in ceiling or basement areas is not too difficult to accomplish. To apply this system to an existing home may be more difficult. It would require exposing piping at the furthest fixture and piping the dedicated return line through ceiling or basement areas. However, in some installations the dedicated line circulation system may be more cost effective than the other methods.

Gravity Water Circulation System

The *gravity* water circulation system uses the natural tendency for hot water to rise and cooler (heavier or denser water) to fall. The system is designed and piped similarly to the dedicated line water circulation system; however, the water heating equipment must be located below the level of the fixtures or appliances served. The hot water will rise out of the heating equipment, through the piping to the highest point of use. A dedicated return line is provided at this point to permit circulation of the cooler water back to the heating equipment by gravity. A pump is not needed because gravity does the work in this system.

The action of the water in this system is relatively slow. If hot water usage is limited to a few times during a day, this system will work very effectively and efficiently if installed correctly. The greatest benefit of the gravity water circulation system is that no energy is used in the movement of water while providing hot water to the fixture when needed.

One of the primary reasons for providing plumbing in a home or building is convenience and comfort. The use of hot water for cleaning and cooking purposes is also one of the reasons that plumbing has helped to advance our lifestyle. However, part of this lifestyle is that we expect our hot water to be instantly hot. Instead of using the water as it comes out of the tap until it becomes hot we wait for the water to warm while the cooler water goes down the drain wasting thousands of gallons of water per year. Water circulation systems help to eliminate the time waiting for water and the resultant squandering of our precious resource.

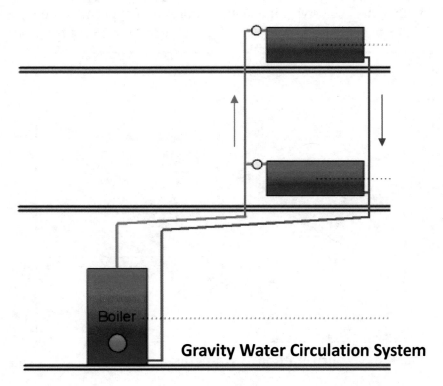

Gravity Water Circulation System

Layout

Grouping

Trunk

Branch

Manifold

Water Distribution Piping Installation

The methods and materials used in the design of the water distribution system can make a great impact on water use and the energy needed to heat water for domestic purposes. All facets of the system must be examined to produce the most efficient system possible. In fact, the entire building should be scrutinized by the architect, design professional or homeowner with energy and water use in mind to create the most sustainable building design. This process begins with the *layout* of the rooms within the building itself.

Designing rooms with plumbing fixtures closer together or on the same side of the building will help to reduce the amount of piping needed for the plumbing system. *Grouping* fixtures in those rooms closer together or for back to back installation will also reduce both drainage and water piping needs. Placement of water heating equipment close to the cold water main supply and adjacent to the area where most of the hot water will be used will also decrease the amount of piping needed. These shorter distances will create shorter lengths of pipe and help to reduce pipe size and the pressure needed to deliver the proper volume of water needed at fixtures and appliances. This reduction in pipe size and pressure will decrease the overall volume of water used or heated and thus save water and energy.

The method used to pipe the water distribution system can also lead to reductions in pipe size and pressure. The conventional method discussed already, the *trunk* and *branch* method, can be improved upon in some instances by using the *manifold* method of piping. In this installation, larger trunk lines are connected to a manifold. Branch lines serving individual fixtures and appliances are piped from the manifold to each fixture and appliance rather than from a long trunk line.

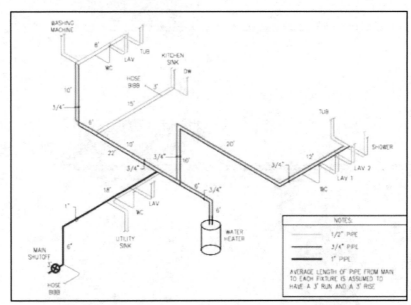

Trunk and Branch Method of Water Distribution Piping.

As shown in the illustrations the manifold system uses more piping than the trunk and branch method, however, the individual fixture supplies are of a smaller size. This can lead to an estimated 65% decrease in the volume of water flowing to the fixture and consequently a 65% decrease in the energy needed to heat that water. If the distribution system is piped without a circulation system, it will also lead to a 65% less loss of water down the drain while waiting for the hot water to arrive.

Copper pipe
PVC
CPVC
PEX

The materials used in the water distribution system can also be a factor in water use and energy reduction. *Copper pipe* or tubing has been used for many years to pipe the water distribution system. It still is a good choice for material especially since it is totally recyclable. However, the increase in cost for copper has led to the use of other materials such as *PVC*, *CPVC* and *PEX* piping. The manifold system is not exclusive to PEX; however, PEX is used in the vast majority of those systems. PEX piping can be a simpler material to use and is readily available in the smaller sizes allowable in the manifold system. Plumbing codes and manufacturer's installation requirements should be closely adhered to and followed.

One common installation practice that should be avoided concerning PEX tubing is the bundling of hot and cold tubing together in ceiling or basement installations. This can lead to heat transfer to the cold water tubing and heat loss in the hot water tubing. This will of course defeat the purpose of using the manifold system and PEX tubing by causing delays in the desired water temperature in both hot and cold water at the fixture.

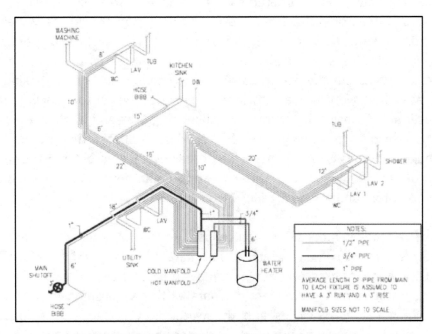

Manifold Method of Water Distribution Piping.

Protection of the Water Distribution System

Installing an efficient and sound water distribution system also requires that it must be protected from damage and *cross-connections* or *backflow*. Damage to the water distribution system can occur from *water hammer* and *thermal expansion* causing leaking faucets, fixtures and appliances while also leading to water heating equipment failure. This will of course lead to losses of potable water and increased energy costs. Cross-connections can lead to the contamination and / or loss of potable water and possible health hazards.

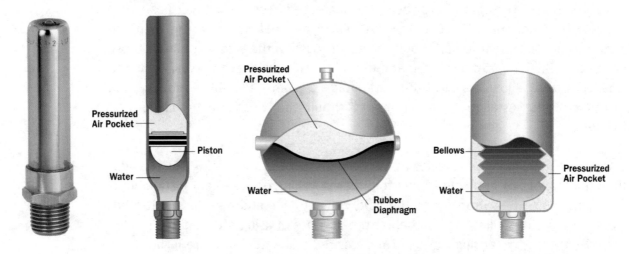

Types of Water Hammer Arresters

Water hammer occurs when the flow of moving water is suddenly stopped by a rapidly closing valve. Flushometer valves, some electronically actuated valves such as in a clothes or dishwasher are examples of rapidly closing valves. The sudden stop in water flow results in a tremendous spike of pressure behind the valve, which acts like a tiny explosion inside the pipe. This pressure spike can reverberate throughout the water distribution system, rattling and shaking pipes, until it is absorbed. Normally, a sufficient pocket of air will absorb such a pressure spike, but if no pocket of air is present, faucets, fixtures and appliances supplied by the system could be damaged as they are left to absorb this pressure spike. The damage will usually be in the form of leaking faucets or appliances, water closets that constantly run and water heating equipment that leak or develop faulty temperature and pressure relief valves.

The most effective means of controlling water hammer is a measured, *compressible cushion* of air, which is permanently separated from the water system. One model of *arrester* shown employs a *pressurized cushion* of air and a two o-ring piston, which permanently separates this air cushion from the water system. When the valve closes and the water flow is suddenly stopped, the pressure spike pushes the piston up the arrester chamber against the pressurized cushion of air. The air cushion in the arrester reacts instantly, absorbing the pressure spike that causes water hammer. Many other styles of water hammer arresters are manufactured. There are residential installation models and industrial models, which are also represented in the illustration.

Always consult the specific manufactures detailed installation literature and product manuals for the proper size and installation location for a water hammer arrester.

Key Words

Compressible cushion
Arrester
Pressurized cushion
Polluted
Contaminated

A cross-connection is any actual or potential connection or structural arrangement between a potable water system and any other source or system through which it is possible to introduce any used water, industrial fluid, gas, or substance other than the intended potable water with which the system is supplied. These cross-connections can lead the backflow (reversed flow) of either *polluted* or *contaminated* water into the potable water system. They can also lead to the flow of potable water out of the system and can remain undetected causing a severe loss of water.

Residential Installation of Backflow Protection Assembly

Polluted water may have an undesirable taste, color, or odor, but it is not considered unfit for human consumption and will not cause sickness. Contaminated water is water that is not safe for human consumption and will cause sickness or even death. In either case, the proper method of backflow protection must be provided against the loss of water and for the protection of the water supply and the health of the occupants of the building. Plumbing codes and water authority regulations should be consulted for proper installation of backflow protection.

Another important key to an efficient and safe water distribution system is maintenance. Fixtures, appliances, faucets and valves must always be in proper working condition. Thousands of gallons of water per year are lost from just one leaking faucet. Multiply this by millions of faucets, fixtures and buildings and the amount of water lost is staggering. Most of this lost water can be saved by simple repairs that anyone can do such as fixing a leaking water closet "flapper" or tightening a valve packing gland. .

Water Heating Equipment

Heating water for domestic purposes is estimated to consume 10 to 25% of the total energy use in an average home. In order to have the most efficient water distribution system it is imperative to choose the correct type and size of water heating equipment that will match the water distribution system and the building application. The selection of the proper water heating equipment must take into consideration type, size, fuel type, overall cost estimates and the equipment's *first hour rating*. An architect or plumbing design professional should be consulted on a commercial or industrial installation as to which type of system to install. The homeowner will usually decide on what type of system to install in a residential setting.

Rather than go through the many different choices that can be made in selecting a water heating system, for our purposes we will simply describe the system and its relevance to green plumbing.

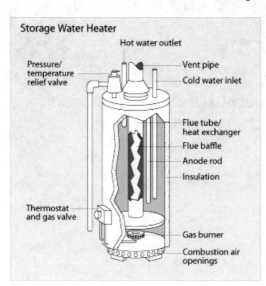

Storage Water Heater

There are five common types of water heating equipment:

- Conventional *storage* water heaters
- *Demand* (tankless or instantaneous) water heaters
- *Heat Pump* water heaters
- *Indirect* water heaters
- Solar water heaters

Conventional Storage water heaters are the most common type of water heater installed. The heating fuel source can be electric, natural gas, fuel oil or propane. All of these operate similarly by heating the water within a tank and storing it there until used.

The major drawback of the storage water heater and other tank type heaters is its standby heat loss. Water is constantly heated to the set temperature of the water heater thermostat even when hot water is not being used. This can waste a significant amount of energy and even heavily insulated tanks will have standby heat losses. Certain types of storage water heaters such as condensing and direct vent water heaters can reduce these losses as can insulating the tank as referred to earlier. Placing the water heater on a timer can also reduce heat loss.

Demand or Tankless Water Heaters

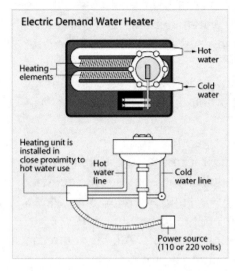

Electric Demand Water Heater

Heating elements — Hot water / Cold water

Heating unit is installed in close proximity to hot water use — Hot water line / Cold water line

Power source (110 or 220 volts)

Tankless water heaters work on the principle of heating water on demand. The heater heats only the water that is used, and continues to heat water until the demand is satisfied. When sized correctly, they can produce an "endless" supply of hot water if needed. These systems have no tanks to store water and as such, do not have stand-by heat losses. Large tankless water heaters are normally natural gas systems requiring venting of the products of combustion. They can provide hot water for large installations by being installed in series creating a large volume of hot water. Electric tankless systems are smaller and are normally for individual fixture use but work on the same principles.

Heating water on demand requires considerably less energy than that required by tank type heaters with stand-by losses. During times when there is low or no hot water demand, tank type water heaters experience heat loss through the insulating jacket of the tank. Most residential water heaters expend energy to heat and maintain water temperature when the house is empty. At times when there is a large demand, tank-type systems may not keep up with the demand. Tankless water heaters solve these problems.

How Does a Tankless Water Heater Work?

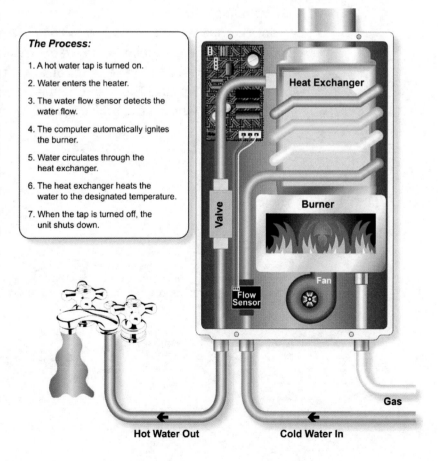

The Process:

1. A hot water tap is turned on.
2. Water enters the heater.
3. The water flow sensor detects the water flow.
4. The computer automatically ignites the burner.
5. Water circulates through the heat exchanger.
6. The heat exchanger heats the water to the designated temperature.
7. When the tap is turned off, the unit shuts down.

Heat Exchanger / Burner / Fan / Valve / Flow Sensor / Gas / Hot Water Out / Cold Water In

Air-source heat pump
Stand-alone heat pump

Heat Pump Water Heaters

Heat pump water heaters are rarely installed in the U.S. Heat pumps are normally associated with heating and cooling systems. However, a heat pump also can be used to heat water—either as a stand-alone water heating system, or as combination water heating and space conditioning system. Heat pump water heaters use electricity to move heat from one place to another instead of generating heat directly. Therefore, they can be two to three times more energy efficient than conventional electric resistance water heaters.

To move the heat, heat pumps work like a refrigerator in reverse. While a refrigerator pulls heat from inside a box and dumps it into the surrounding room, a stand-alone *air-source heat pump* water heater pulls heat from the surrounding air and transfers it dumps it—at a higher temperature—into a tank to heat water. A *stand-alone heat pump* water heating system is an integrated unit with a built-in water storage tank and back-up resistance heating elements. A heat pump can also be retrofitted to work with an existing conventional storage water heater. They require installation in locations that remain in the 40º–90ºF (4.4º–32.2ºC) range year-round and provide at least 1,000 cubic feet (28.3 cubic meters) of air space around the water heater. Cool exhaust air can be exhausted to the room or outdoors.

Heat pump water heaters will not operate efficiently in a cold area, and tend to cool the spaces that they are occupy. An air-source heat pump system that combines heating, cooling, and water heating, called combination systems, pull their heat indoors from the outdoor air in the winter and from the indoor air in the summer. Because they remove heat from the air, this type of air-source heat pump system works more efficiently in a warm climate.

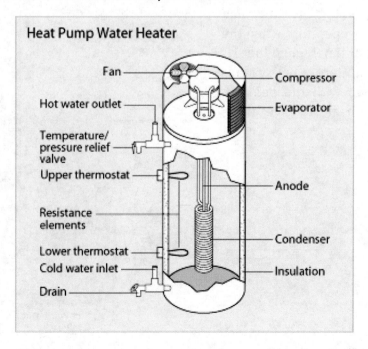

Heat Pump Water Heater

Fan — Compressor

Hot water outlet — Evaporator

Temperature/pressure relief valve

Upper thermostat — Anode

Resistance elements

Lower thermostat — Condenser

Cold water inlet — Insulation

Drain

Indirect Water Heaters

Indirect water heaters work in combination with a building's comfort heating system. The indirect water heater uses the main furnace or boiler heat source to heat a fluid that is circulated through a heat exchanger within the indirect heater's storage tank. The system is an indirect system because the heat source is not located within the storage tank; rather it is located in the furnace or boiler. They can work with *forced air, hydronic* or *radiant* heating systems and with any type of fuel source.

The indirect water heater can be an efficient source of hot water especially for a commercial or industrial installation. The energy used for comfort heating is actually doing double duty by providing for hot water generation. A separate energy source for hot water heating is not necessary providing energy savings. One drawback of this system is the fact that the system must run even in spring or summer months when comfort heating may not be necessary. Unless the installation is in a setting that requires year round comfort heating, this could cause increased energy usage.

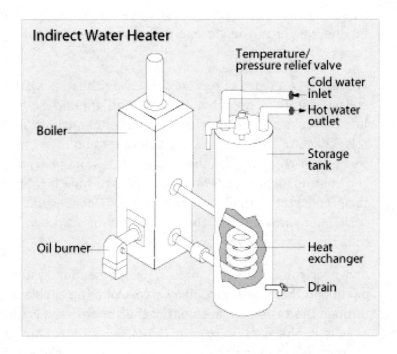

Key Words

Solar collector
Passive
Active
Batch tank
Thermo siphon

Solar Water Heaters

Solar water heaters use the energy from the sun to heat water for domestic water use. These systems can provide the most efficient use of energy for heating water. Free energy from the sun heats water which is then stored in a tank. Water circulates through the **solar collector** by natural means or by use of an efficient pump. Even more efficiency can be obtained by powering the pump with Photo Voltaic solar energy.

A solar water heating system will consist of solar collectors, a circulating piping system, a storage tank and a back up heating system in case of numerous cloudy days. The storage tank and backup system could be a conventional storage water heater designed for this purpose.

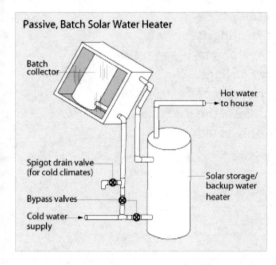

Passive, Batch Solar Water Heater

Batch collector

Spigot drain valve (for cold climates)

Bypass valves

Cold water supply

Hot water to house

Solar storage/ backup water heater

Solar water heating systems are either **passive** or **active** systems. Passive systems work much as the gravity circulating system works using the natural forces of gravity and convection to provide circulation of water through the solar collectors and to the storage tank. However, in this system the storage tank or **batch tank** should be placed above the solar collector so that the water heated by the collector rises into the tank. Care must be taken to account for the weight of the storage tank either on the roof or on upper floors.

There are two types of passive solar water heating systems Integral collector-storage or batch systems and **thermo siphon** systems. Integral collector systems use a tank either within or above the solar collector to heat and store the water and then use gravity and convection to distribute that water to a storage tank or fixtures. The thermo siphon system uses a storage tank located above the solar collector, usually on the roof. Once again, gravity and convection provides for the movement of water from the solar collector to the storage tank.

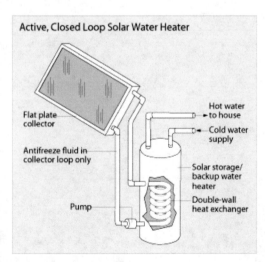

Active, Closed Loop Solar Water Heater

Flat plate collector

Antifreeze fluid in collector loop only

Pump

Hot water to house

Cold water supply

Solar storage/ backup water heater

Double-wall heat exchanger

Active or forced circulation solar water heating systems utilize a pump and thermostatic controls to provide the circulation of water through the system. Placement of the storage tank is not as critical as it is in the passive system. Active systems can be direct or indirect closed loop circulation systems. Direct circulation systems circulate potable water through a solar collector and then to the storage tank which supplies hot water to distribution piping for use. In other words, this system directly heats the water to be used. Indirect circulation systems circulate a heat transfer fluid through the solar collector to a heat exchanger, usually within the storage tank, which in turn heats potable water for use.

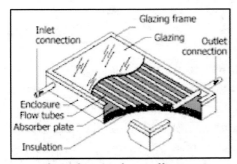

Flat-plate Solar Collector

Integral-storage
Flat-plate
Evacuated tube
Collectors

There are three types of solar collector systems used in solar water heating systems – *integral-storage collectors, flat-plate collectors and evacuated tube collectors*. Integral collector-storage systems have already been discussed. Flat-plate collector systems are the most common systems installed in residential applications. A typical flat plate collector is a metal box with a glass or plastic cover on top and a dark colored absorber plate on the bottom. The sides and bottom of the collector are usually insulated to minimize heat loss. Sunlight passes through the glass and strikes the plate, which collects heat. This heat is then transferred to either potable water or transfer fluid flowing through piping or tubing attached to the absorber plate.

Evacuated tube collectors are more commonly used in commercial or industrial applications. They incorporate parallel rows of glass tubes. Each tube contains a glass outer tube and a metal absorber tube attached to a fin. The fin's coating absorbs solar energy but minimizes heat loss. Water or transfer fluid flows through the absorber tubes collecting the solar energy and heating the water or fluid which is then stored in the tank.

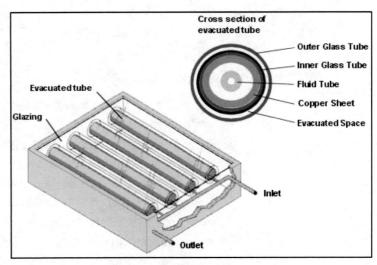

Evacuated Tube Solar Collector

Selecting which type of solar water heating system to use will depend on the type of building, cost, area average temperatures and average amount of sunlight available. The load or amount of hot water needed will also be a determining factor as to which type of system and which type of collector to use, and whether the system should be active or passive. In all cases, plumbing codes, local regulations, and manufacture's installation instructions should be followed.

Solar water heating is one of the easiest and most efficient uses of solar energy. These systems are able to pay for themselves in a few years (depending on costs and usage) and have low maintenance requirements. Solar water heating system components can last for 20 years or more under normal operation. Solar water heating systems, when used in conjunction with traditional systems reduce the electrical and / or fossil fuel requirements for water heating.

Key Words

First hour
ENERGY GUIDE

First Hour Rating

Choosing which type of water heating system to use can be a daunting task. All aspects of the building and the plumbing system – building type, size, fixtures, piping materials, method of installation and layout - must be taken into consideration. In addition, the load or the amount of hot water needed per day must be known in order to select the correct capacity of the system. If the water heating equipment is too large, energy will be wasted in heating unneeded water. If the system is too small, the system will have to work harder and longer to provide the amount of hot water necessary which will also waste energy.

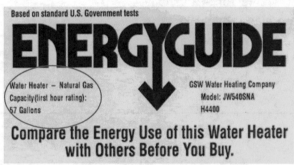

Water Heater Energy Guide Label

The capacity of the water heating system is determined by the first hour rating of the system. The **first hour** rating represents how much hot water the unit can supply in a one-hour period if it starts with a full tank of hot water. The first hour rating is found on the **energy guide** label on the water heater. This will also help to find the most efficient water heater for the particular use.

Plumbing codes provide general rules for determining the proper capacity of water heating systems especially for residential applications. Each manufacturer of water heating equipment also provides methods for determining the correct load and capacity of water heater to use. The U.S. Department of Energy provides a method for determining the load and proper capacity for the system on their Energy Efficiency and Renewable Energy website www.eere.energy.gov. This site also provides information on cost analysis, energy efficiency and other subjects that can assist in selecting an efficient and sustainable water heating system.

Landscape Irrigation Systems

An integral part of a residential or commercial building installation is its landscape *irrigation* system. Most irrigation systems utilize fresh potable water equaling nearly 30% of the residential potable water use. Much of this irrigation is wasted through overwatering. Reduction to the volume of wasted water can be achieved through the careful selection and installation of a landscape watering system.

Water intensive turf grass has been a major landscaping feature in the U.S. until the last decade. Green or sustainable landscaping has been coming into prominence with new principles of *Xeriscaping* or natural landscaping. These landscape methods utilize indigenous, drought-tolerant plants and turfs that when established can eliminate the need for watering. This has lead to significant reductions in the need for water for irrigation.

Sprinkler water runoff

Where these methods of landscaping are not used or where irrigation is necessary better methods of irrigation can also lead to reductions in water usage. The use of sprinklers for watering purposes wastes water by placing water on the top of leaves or grasses rather than where it is needed, the plant's roots. More than 30% of sprinkled water evaporates before permeating the soil and another high percentage of that water runs off the intended site to sidewalks or streets. Irrigation systems that place water in the soil must be used to eliminate these losses.

In ground or *drip irrigation* systems should be used that place the water where needed rather than on top of the soil. These systems can be total underground irrigations systems and/or utilize drip emitters at the plant to distribute water. Further water use reduction can be accomplished by using timing systems that also incorporate weather-tracking systems that will energize the system only when necessary and when rainfall is not present.

These systems can be augmented by substituting potable water with water from rainwater harvesting systems, gray water systems, or reclaimed water.

Drip irrigation emitter

Another important part of conserving water is the maintenance of these systems. If a sprinkler head or emitter becomes defective or is dislodged a significant amount of water can be lost until repaired. Great losses will also occur if piping or valves leak. Whether it is commercial or residential landscaping, inspection and maintenance of the irrigation system should be included with the plant or turf maintenance schedule.

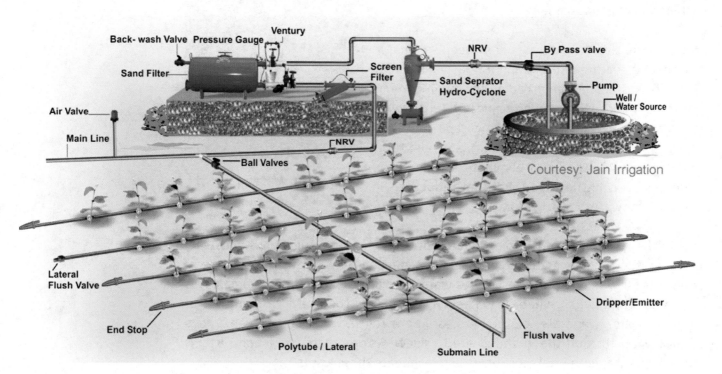

Typical In-ground Irrigation System

Waste Water Re-use Systems

Utilizing wastewater is one more way of perpetuating sustainability or green building. Capturing and re-using the hundreds of gallons of water used in the daily building hydrologic cycle *before* it enters the sewer system will not only conserve this resource but ease the burden placed on sewage treatment facilities by this wastewater. This alone will save billions of dollars in energy and facility costs by lessening the need for larger treatment systems and the energy used to treat this wastewater. It will also reduce the amount of waste water that will be discharged into our potable water supply sources - rivers, lakes and oceans – and thus keep these precious resources from possible contamination.

Unlike the conservation of potable water, the re-use of wastewater usually entails more complicated piping installations than can be accomplished by a homeowner or maintenance personnel. Retro fitting an existing building to utilize the systems outlined below may include some demolition of walls and excavation work. It is far easier to include these systems in new construction.

Wastewater re-use systems include the following systems that capture heat from wastewater or capture the wastewater for, reuse in the building, or for irrigation, or for private wastewater disposal:

♦ Drain water heat recovery systems

♦ Gray water systems

♦ Reclaimed water systems

♦ On site waste water treatment systems

Green Awareness

Drain Water Heat Recovery (DHR)

Water used in showers, lavatories, clothes washers, and dishwashers is normally hot water. After use, this hot water goes down the drain taking its energy in the form of heat with it. This represents typically 80-90% of the energy used to heat water.

Preheat

Coiled heat exchanger

Drain water heat recovery systems provide a method to capture this lost heat before the wastewater leaves the building or is diverted for treatment.

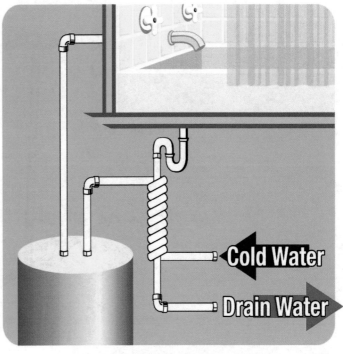

DHR systems use hot wastewater as it flows through the drainage system to **preheat** cold water for use in water heating applications. The system consists of a specialized **coiled heat exchanger** mounted to or replacing a section of a drainage pipe. The heat exchanger is connected between the incoming cold water supply and the water heater cold water inlet.   Heated wastewater flowing down the drainage pipe pre-heats the cold water flowing through the heat exchanger on its way to the water heater, effectively capturing most of the wastewater heat.

DHR Single Fixture System

There are two common types of heat exchangers utilized in drain water heat recovery systems. The first uses only one source of drain water such as a bathtub waste. The water supply outlet connection is then connected to the inlet water supply of the water heater. This relatively small installation can be connected directly to the hot water supply of a fixture itself, preheating only the fixture hot water instead of the water heater. The second type of heat exchanger is connected to or replaces a segment of the drainage piping containing flows from multiple hot water supplied fixtures. The exchanger is connected to the inlet side of the water heater preheating the cold water entering the heater.

These systems preheat the cold water supply to a fixture or water heater or they can be connected to heat a storage tank. The storage tank is connected to the inlet of the water heater. The use of a storage tank allows recovered heat to be stored rather than rely only on the simultaneous use and preheat of the cold water supply without a storage tank. The addition of a storage system will also allow for the preheating of water in "batch" use such as for baths or clothes washing.

It has been estimated that up to 40% of the energy needed for domestic water heating can be saved by using these systems. These reductions will vary and will depend on the type and number of fixtures connected to the system and to the type of DHR system itself.

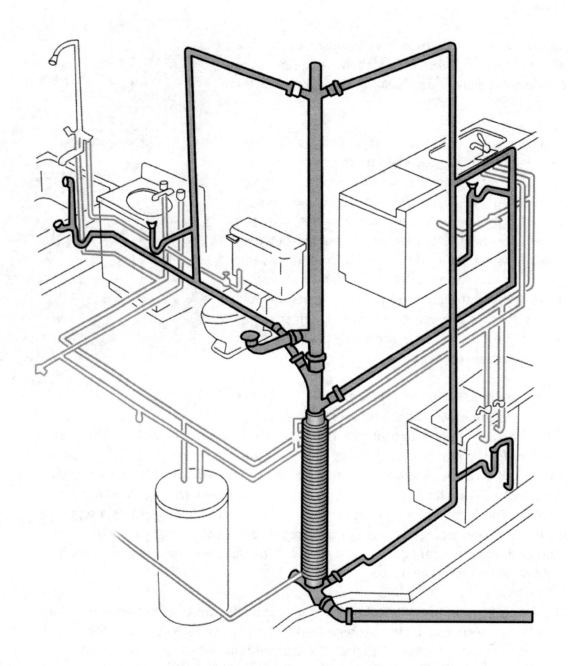

DHR Multiple Fixture System

Gray Water and Reclaimed Water Reuse Systems

Key Words

Gray water

Black water

Reclaimed water

Very little of the potable water used within a home or building is consumed as drinking water by the occupants of these structures. Estimates vary but the World Health Organization suggests that .5 to 1.0 gallon per person per day is needed for drinking and another gallon for cooking and food preparation. The average American uses approximately 100 gallons of potable water per day. The rest of this potable water, approximately 95-98 gallons per person per day, is used for various purposes, discharged into fixtures and ultimately flows through sewers to sewage treatment facilities and through the rest of the hydrologic cycle. The creation of a sustainable environment and the construction of green buildings must include capturing of this wastewater and reusing it.

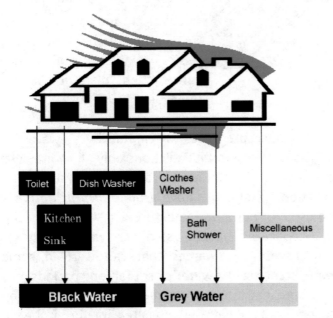

Identification of Black Water and Gray Water

Wastewater discharged from fixtures within buildings is classified into one of two categories – **gray water** and **black water**. Gray water is wastewater from bathtubs, showers, lavatories, wash basins and clothes washers. Specifically, "Gray water flows through fixtures that do not come in contact with fecal matter or food solids. Black water is wastewater from fixtures that do come in contact with or contain fecal matter or food solids such as water closets, urinals, kitchen sinks or dishwashers. It should be noted that if a clothes washer is used for washing diapers, then the clothes washer wastewater could contain fecal matter and should be classified as black water.

The wastewater that is captured from gray water and black water fixtures can be utilized in two distinct reuse systems – gray water systems and **reclaimed water** systems.

Gray Water Systems

Single Fixture Use Gray Water System

It is estimated that 60% of water used within a home or building becomes gray water. Capturing this gray water for reuse in irrigation systems or for fixture use is the purpose of gray water reuse systems. Depending upon the system employed the captured gray water can be separated from the black water waste and used with no other treatment necessary other than filtering solids. In some cases, it is necessary to re-pipe an existing plumbing system to keep black and grey water separate. The installation of this type of system is much easier when included in the initial construction plan for a new home or building.

Key Words
———————

Subsurface

Drip irrigation

Pathogens

Cross Connection

There are currently two accepted uses for captured gray water in and around a home or building. Many codes and jurisdictions only allow gray water to be used for irrigation systems. The majority of these systems are *subsurface* (underground) irrigation systems or above ground *drip irrigation* systems. Gray water may also be used for non-potable fixture use such as flushing water closets or urinals. There are concerns that the use of gray water for these purposes may expose humans to harmful *Pathogens* that can exist in some gray water. Untreated gray water standing in a water closet for long periods may create odors. These situations can present human health problems. Health departments and local plumbing codes should always be consulted for the proper installation and use of gray water systems.

Small packaged single fixture gray water systems are available, although, most of these systems are not yet accepted by any of the model plumbing codes. These systems capture the wastewater from a bathtub or lavatory, store the gray water in a small storage tank and utilize it for flushing the water closet. When the storage tank is full, the wastewater is diverted into the drainage system. This is a simple and relatively easy system to install. Extreme care must be taken to prevent *cross connection* with the potable water system. The local health department, plumbing codes and authority having jurisdiction over plumbing systems should again be consulted for approved use and installation of these systems.

Multiple fixture use systems are much more complicated piping systems than the single use system. Simply put, the multiple fixture gray water reuse system separates the plumbing drainage system into two systems. Black water fixtures are piped normally and connected to the building drain or sewer system. Gray water fixtures are piped to a separated drainage system, which connects to a gray water storage tank. This tank in turn connects to an irrigation system that will then utilize the gray water. The gray water storage tank overflow is connected to the building drain or sewer system so that unused gray water does not remain in the tank for long periods. Gray water systems can be small planter systems or very large turf and shrubbery systems. They are available as package units or can be piped as in the illustration. Each installation and use of gray water will determine the extent and type of system to be used.

The gray water system does not have to connect to all gray water fixtures or utilize all of the gray water created within a building. The system should be sized to utilize enough of the gray water to provide for adequate irrigation of the building property. What is not wanted is the *runoff* or *ponding* of gray water on the site. This could lead to the possible health problems. Most model plumbing codes have

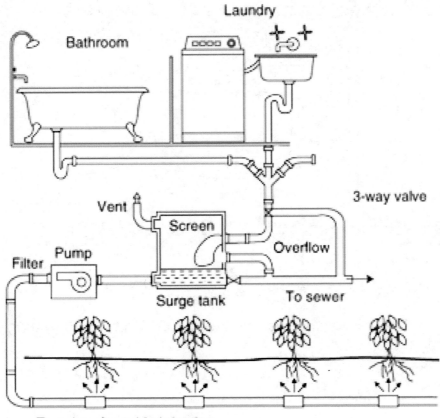

Multiple Fixture Gray Water Piping System

specific sizing requirements for gray water irrigation systems.

One other safety point – some of the pathogens that exist in gray water may be transferred to some plants. Therefore, gray water is not recommended for plants and fruits that may be used for human consumption. Nor is it recommended that gray water be used in above ground sprinkler irrigation systems, as the gray water will lie on the surface ground rather than below ground. If gray water is used for these purposes or for non-potable fixture use it should be treated to the level of reclaimed water.

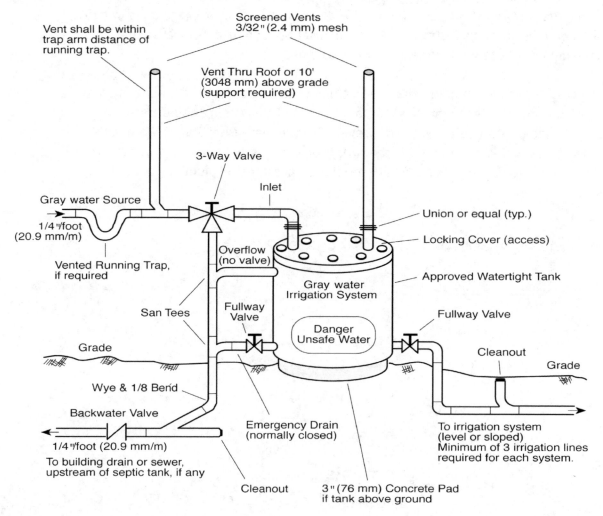

Gray Water Holding Tank and Piping – Gravity System, From 2006 Uniform Plumbing Code

Reclaimed Water Systems

Reclaimed water systems take wastewater reuse to the next level. These systems utilize all of the wastewater, gray and black water, and treat this wastewater to specified levels and then return the resulting reclaimed water for use in irrigation systems and for non-potable fixture use specifically water closets and urinals. Usually these systems are built by the municipal water authority under public works projects, which utilize the sanitary waste from a section of a metropolitan area or from the entire area. However, recently smaller development or single building package systems have been installed. These systems normally have been installed only for commercial or industrial use but they are currently being used in some areas of the country for residential use.

Reclaimed water

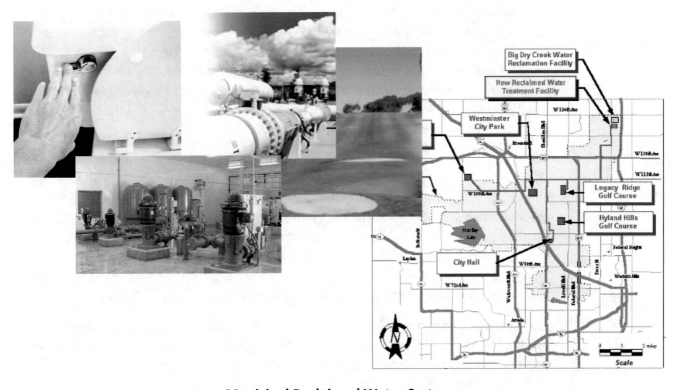

Municipal Reclaimed Water Systems

Reclaimed water is not potable water. It is defined in the 2006 Uniform Plumbing Code as:

"Water that, as a result of tertiary treatment of domestic wastewater by a public agency, is suitable for a direct beneficial use or a controlled use that would not otherwise occur. The level of treatment and quality of the reclaimed water shall be approved by the public health Authority Having Jurisdiction."

The code further states that:

"tertiary treatment shall result in water that is adequately oxidized, clarified, coagulated, filtered, and disinfected so that at some location in the treatment process, the seven (7) day median number of total coliform bacteria in daily samples does not exceed two and two-tenths (2.2) per one hundred (100) milliliters, and the number of total coliform bacteria does not exceed twenty-three (23) per one hundred (100) milliliters in any sample. The water shall be filtered so that the daily average turbidity does not exceed two (2) turbidity units upstream from the disinfection process."

Essentially reclaimed water is water that after *treatment* and *disinfection* will not cause disease if contacted or ingested but it is not treated to final potable water levels.

A reclaimed water system does not affect the internal building drainage system. Sanitary waste from the building, buildings or sewage systems is captured at a remote treatment facility. The reclaimed water is piped from that facility to the point of use whether it is for onsite irrigation or for non-potable fixture use. the potable water supply system is affected in that there must be clear identification of all potable water and reclaimed water piping and systems whether inside the building or in the ground on or off the property.

Reclaimed water is required by most codes to be run in purple pipe and identified as "Reclaimed Water – Do Not Drink". Care must also be taken that the reclaimed water and potable water system is not cross-connected even with the use of a backflow prevention device. The 2006 Uniform Plumbing Code requires that where reclaimed water is used for water closets and urinals, the potable and non-potable piping be kept separate within the building. Although reclaimed water will not cause immediate sickness, long term ingestion of reclaimed water will cause sickness therefore it is important that the local jurisdiction requirements be followed.

Reclaimed water is normally associated with golf course irrigation but it is being used for far more than just golf courses. Municipal and industrial reclaimed water use now accounts for more than 870 billion gallons per day. Along with golf course irrigation, its uses include certain agricultural irrigation, decorative features such as fountains and ponds, cooling tower water makeup, concrete mixing, snowmaking, fire protection lines and fire sprinkler systems, and non-potable fixture use.

♦ Using gray water and / or reclaimed water will:
♦ Reduce potable water consumption.
♦ Reduce the impact on septic and sewer systems and treatment facilities.
♦ Reduce the energy used to pump and process water.
♦ Recharge the ground water.
♦ Increase the nutrient value of top soil.
♦ Stimulate plant growth.

The ultimate goal in the reuse of sewage or wastewater is the total recycling of that water bypassing the last stages of the hydrologic cycle - discharge to rivers and streams. In these systems, the sewage wastewater is treated to potable drinking water levels and returned to the water supply system. However, these systems are not being developed in large numbers primarily because of the *"yuck" factor*. The unknowing public just does not want to think that yesterday's water that flushed the water closet is the same water being used to brush their teeth today.

Key Words

Yuck factor

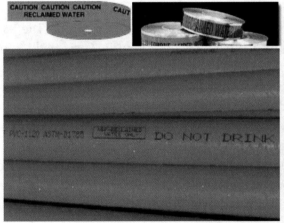

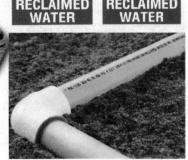

Reclaimed water piping, marking tape, underground marker tape and signage

Onsite Wastewater Treatment Systems

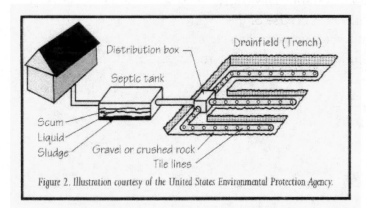

Figure 2. Illustration courtesy of the United States Environmental Protection Agency.

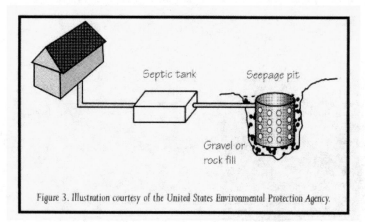

Figure 3. Illustration courtesy of the United States Environmental Protection Agency.

Private Sewage Disposal Systems

Onsite wastewater treatment systems do not necessarily reuse the building wastewater rather they dispose of this wastewater without sending it to a sewage treatment facility eliminating the energy cost of transport and treatment. The most common is the private sewage disposal system or the *septic tank* system. Other systems utilize nature either directly or indirectly to accomplish wastewater treatment.

The private sewage disposal system or the septic tank system captures waste from a house or building in a septic tank. The septic tank settles out the solids, allows fats, oils and grease to float on top of a clarified liquid that eventually flows out of the tank by gravity into subsurface disposal fields and or seepage pits. The liquid or effluent is absorbed into the ground. The use of these systems is very land intensive and is used only for less densely populated areas. The system must be designed properly otherwise the disposal area will become saturated and unable to absorb and treat further effluent and may contaminate the ground water table. This is the very reason that in many areas these systems were replaced with sewer systems and sewage treatment facilities.

Key Words

Septic tank

Wetlands

Highly efficient buildings are being designed to dispose of their waste through the direct or indirect use of nature. The direct use of nature uses *wetlands* either natural or constructed to treat the effluent from the building. These wetlands are ecological systems that can break down organic waste from the effluent using the remaining nutrients to sustain the wetlands.

The constructed wetlands can either be surface or subsurface wetlands. The surface system consists of shallow basins that allow the wastewater to flow slowly from basin to basin eventually clarifying the wastewater, which is released as clear water. The subsurface system allows the waste to flow through a substrate such as sand and gravel clarifying the wastewater and eventually releasing it as clear water. The subsurface system works much better in cold climates and eliminates much of the possible contact of effluent to humans.

As one can imagine these systems are very intricate and labor and land intensive. They will normally be used for those large installations seeking the greatest extent of ecological design. The costs of construction of the wetland versus the cost of waste treatment should also be considered.

Nutrients

Bacteria

The indirect use of nature as wetlands brings these wetlands into the building itself. A constructed wetland called a living machine contains biological organisms that breakdown wastewater components into nutrients that are then fed into a constructed wetland inside or outside the building. They can be tailored to the specific waste **nutrients** that exist in the effluent and the constructed wetland can be easily expanded. Essentially the waste flows through a series of plant, **bacteria** and even fish tanks while providing a garden like atmosphere.

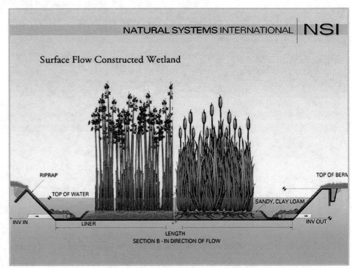

Any one of these wastewater reuse systems will provide for a reduction in the amount of freshwater use, energy consumption and environmental contamination. In combination, these systems can almost totally eliminate the disposal of wastewater to treatment facilities and dramatically reduce the amount of water used in the modern building. The only caveat is that most of these systems must be maintained in good order and not left to themselves. We do not need contaminated groundwater, wetlands, and freshwater resources that need remediation.

In addition to the above wastewater reuse systems there are other systems within the building, which develop wastewater that can be captured for use.

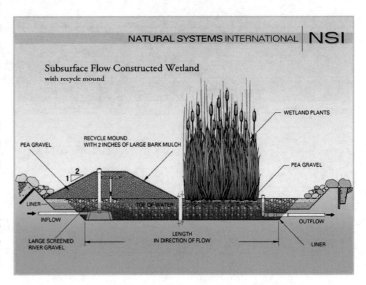

HVAC systems create water condensate that normally flows to the drainage system. The capture and reuse of this type of wastewater can be included in gray water systems quite easily.

Swimming pool backwash systems normally discharge to the sewer or to the ground wasting good water. Systems that return this backwash water for use in the pool are available and are relatively easy to install.

Water treatment systems, especially the type that backwash and drain to the sewer are often misused. The backwash or regeneration settings are set too frequently and waste water. In fact, some areas of the country have prohibited their use. The use of a either a whole house filter or at least an activated carbon filter at the source can eliminate these systems and provide the treatment of drinking water without the waste.

Facilities that use large amounts of water such as car washes are required in some areas to recycle this water through systems much like gray water systems rather than let the wash water flow down the sewer.

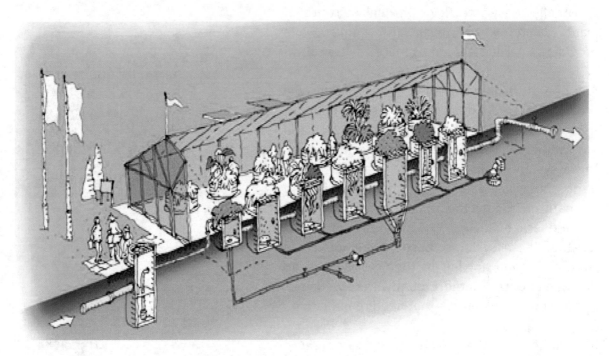

Diagram of a Living Machine

Rainwater Harvesting

Rainwater, the foundation of the hydrologic cycle, replenishes our natural sources of freshwater, rivers and lakes. It is only natural to attempt to utilize this abundant (in areas without drought) resource. Capturing this natural source of freshwater has become another element in green plumbing systems and sustainable construction design. After all, rainwater retention was one of the original methods of providing water for domestic use. The rain barrel catching rainfall from the roof was a feature of the home before written history. It is only fitting that plumbers now come back to the origins of our trade. The fact of the matter is that effective catchment and use of rainwater can further reduce the need for freshwater especially for such wasteful uses as decorative lawn and shrub irrigation. Many of the methods of harvesting rainwater can be simple installations or can be implemented in existing storm water systems.

Rainwater *harvesting* systems can be as uncomplicated as a rain barrel under a downspout with a spigot at the bottom to attach a garden hose. In fact, this can be a very effective irrigation supplement. Several package systems are available that can be integrated into a lawn and shrub irrigation system. These systems efficiently use the capture rainwater and consist of a storage tank, pumps and electric controls. The controls to the system will sense the lack of rainwater in the storage tank and switch on the fresh water system when there is a lack of rainfall.

Rainwater harvesting systems can also be large elaborate systems. Rainwater captured on or from roofs is piped and contained in large cisterns for storage. These systems often have roof washing systems that remove the initial flows from the roof that will normally contain dust or other particulates and then divert the rest of the rainwater into the cistern. They will also employ a filtering system to remove leaves or other particles that may plug valves or irrigation emitters. Pumps will provide conveyance of the rainwater to the point of use unless the cistern or storage tank is situated high enough to allow gravity drainage. A control system will integrate the fresh water and rainwater systems to provide for the most efficient use of water.

Existing roof and storm water drainage systems can easily be utilized for rainwater harvesting. Roof drains and storm drainage piping already exist in most large buildings and can be piped to connect to storage areas. Certain types of roof systems can increase the ability to capture rainwater such as the controlled flow roof system. This is an engineered system that captures rainwater on the roof and temporarily stores it there by controlling the flow through the storm drainage system at a predetermined rate. This type of system can very easily be adapted to a rainwater harvesting system.

Other sources of rainwater are paved or graded areas for parking or other uses surrounding a building that normally are sloped to drain from the building site. Careful planning can utilize these areas for rainwater catchment that can also be used for irrigation or other non-potable water uses. In fact, in many areas of the country, especially surrounding ocean or beach areas, the property owner is obligated to capture this water and pre-treat it to remove oils or other contaminants before discharge from the site. These systems will operate similar to sand and oil separation systems and can easily be adapted for rainwater harvesting.

Water captured through these rainwater harvesting systems should be filtered even for use in irrigation. Roof systems that utilize asbestos or asphalt materials should not be used for these purposes because of the possibility of contamination. In some installations, the captured water is used for non-potable fixture use such as water closets. This will entail further water treatment similar to reclaimed water systems to be used within the building. One other caveat concerning the installation of these systems is that there are no clear standards or code regulations for rainwater harvesting systems. Until the development of new standards, each system will be unique and care should be taken to ensure proper installation and protection of the potable water system and the occupants of the building.

Fire Protection Systems and the Environment

Another portion of fresh water use that must be addressed in a sustainable green building is the fire protection system. Fire protection systems are divided into three main categories; Industrial, Residential, and the Water Supplies for both.

House Fire and Resulting Debris

Industrial Fire Protection Systems:

Industrial fire protection systems are installed in various applications such as, hotels, nursing homes, dormitories, hospitals and retail office buildings and also chemical, manufacturing and power generation facilities to name a few. These systems range from water based to chemical based systems that use foam or gaseous mixtures to control fire. As matter of fact the gaseous fire protection systems using Halon 1301 was one of the first examples of environmental or green concerns.

Halon 1301 was developed for use in water sensitive areas such as computer and telecommunication facilities. The ***inert, non-conductive*** and safe gas was ideal for these uses however, it was later found to be one of the most harmful compounds to the Ozone layer of earth's atmosphere. Different compounds and systems were developed to replace Halon 1301and protect the Ozone. Beginning in 1993, FM 200 a gaseous clean agent with no such environmental impact and water mist systems were installed in place of Halon systems.

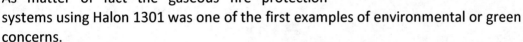

Key Words

Inert

Non-conductive

The water mist system is also being used to replace conventional deluge wet pipe systems that use large amounts of potable water to extinguish fires. Water mist systems apply a very fine mist of water that can quickly suppress a fire. This can reduce the amount of water used by the deluge system by as much as 50 to 80 percent. The mist system can also reduce the size and amount of piping used in the fire protection system in comparison with a conventional wet system. In a typical application, a

Water Mist Sprinkler Head

conventional deluge system might use a 6 inch pipe system, while a water mist system would use 1 inch piping and if activated during a fire would only flow 1 to 2 gallons of water per minute while the conventional system would use 10 to 40 gallons per minute. Water mist systems can reduce the amount of water used in a potential fire and reduce the actual amount of piping material and cost of the fire protection system while limiting the damage that can occur using the water deluge system.

Residential Fire Protection Systems:

**Residential Fire Sprinkler
Connected to Domestic Water System**

The benefits to industrial fire protection systems can also be applied to residential fire protections systems. The fact that the installation of fire protection systems in homes is seen as an add-on feature rather than a mandatory requirement needs to be addressed. We know that fire protection systems can save lives but they can also reduce the amount of waste created by damaged or destroyed homes and eliminate the amount of toxins released in the air through smoke which can harm the environment or create health problems. These systems should be an essential part of any green home.

Residential fire protection systems where installed can utilize the existing potable water system rather than creating another piping system within the house. A multi-purpose pipe system, which shares the domestic water system, can reduce the amount of piping needed to supply a residential fire protection system. Include this with a mist system and the reduction in water, material use, and potential damage can be quite significant.

Water Supply:

Fire protection system water supplies are normally potable water lines connected to the same water distribution system as the building water supply. Therefore, these fire protection systems can also utilize *harvested* rainwater and *reclaimed* water systems, which can increase the benefit of these systems. Normally these fire protection systems are supplied using pumping systems that use significant amounts of energy. Water supplies that are fed from gravity tanks instead of pressure tanks can reduce energy use by eliminating the need for large pumping systems, which use a great amount of energy to operate.

Another method to reduce water consumption occurs during the required yearly testing of the fire protection system. The water used during yearly testing cycles, for example alarm flow test, drain test and annual standpipe testing, can be recaptured instead discharging this water down the drainage system or the street. Water storage tanks can be located to receive any water discharged during this testing and be used for test water or supplemental water for fire protection.

Instead of fighting a fire by responding fire departments, the use of a fire protection system when activated by a fire will in create a reduction in the use of water. A typical fire will activate one or two sprinkler heads equating to around 346 gallons of water used per fire while containing and or extinguishing the fire. In comparison, the typical fire department would use 2,935 gallons of water discharged from their fire hoses. Fire sprinklers will thus reduce the amount of runoff from the structure and possibly into the ground water or nearby streams, lakes, or other large bodies of water. Runoff from the use of fire hoses could contain chemicals or other toxins stored at the site, which can cause contamination of the surrounding areas.

It is also noteworthy to mention the reduction of construction waste by the use of fire sprinklers in buildings. Typically, after a structure fire where fire sprinklers are used there is limited amount of construction debris, if any. Debris from a partial or total loss of a structure without sprinklers could be very great. It is reported that construction and demolition debris represents up to 21 percent of landfill debris. Construction materials needed to repair or replace a damaged or destroyed building must also be taken into consideration for this takes energy and resources to manufacture, distribute, and reconstruct such buildings.

Fire sprinklers and their quick control or extinguishment of burning materials can eliminate a great deal of this unneeded waste and work.   Indeed fire protection systems are designed to protect property and life and they can be designed and installed to protect our water resources and the environment too.

Key Words

Harvested

Reclaimed

LEED

Green Plumbing System Relevance to LEED

The practical application of the Green Plumbing System will result in significant reduction in the use of potable water, wastewater treatment and energy savings. It can also result in credits in the *LEED* Building Assessment System. Up to 5 total credits can be achieved in the LEED-NC Water Efficiency Category.

Reducing potable water consumption use for landscaping by 50% can result in 1 credit. (WEc 1.1)

The total elimination of potable water for landscaping can result in 2 credits. (WEc 1.2)

- ◆ Use of gray water or rainwater in place of potable water is allowable for both the 50% reduction and elimination credits.

Innovative wastewater technologies can result in 1 credit through strategies that reduce building water consumption and thus reduce wastewater flow by the following two methods (WEc 2):

- ◆ Reducing potable water used for sewage conveyance by at least 50%
- ◆ Treat at least 50% of the water to tertiary standards on-site and then use the waster on site or infiltrate it back into the ground.

Water use reduction can result in up to 2 credits (WEc 3.1 & 3.2):

- ◆ One credit can be achieved by reducing potable water use in the building by 20%
- ◆ An additional credit can be achieved by a reduction in potable water use by 30%.

Other LEED credits can be achieved for onsite use of renewable energy sources. The use of solar water heaters could garner as much as 3 credits if it provides 12.5% or more of the total building energy use (EAc2).

High Efficiency Building Design is becoming the norm these days rather than the exception. The workforce constructing these buildings and homes must be aware of these systems and their relevance to LEED to be affective in this market. Although the actual assembly and installation of the plumbing system is relatively the same whether green systems are used or not, the service plumber, new construction plumber and the apprentice plumber can be very influential in the design of the plumbing system. It is also important to contractors that their workforce be certified as knowledgeable about green building design. This can be critical to the contractor in winning a bid on a LEED certified building.

Summary

Although most of the Green Building Design and the LEED Building Assessment System seems concerned with energy performance, reduction in water use is also critical to the sustainability of our planet. If we do not conserve and wisely use this precious resource and dispose of our waste without contaminating that resource our way of life will surely be in jeopardy.

The green plumbing systems discussed in this manual can make great inroads in the protection of our water resources. If these systems are employed, we can reduce the 100 gallons of water used per person per day to a more manageable 25 gallons or less. However, if the infrastructure of our water supply system is not addressed we will be condemning humankind to a bleak future.

It is estimated that 80% of fresh water consumption is associated with agricultural irrigation and that 60% of this irrigation water is wasted due to leaks and mismanagement. In many large and consequently old cities throughout the world, the potable water supply systems are estimated to waste 40% or more of the supplied water. In most instances, the original water supply systems, installed over 100 years ago, are still in use. These systems contain countless leaks, causing damage to buildings and infrastructure. These problems must be corrected if our society wishes to be truly Green. Investment must be made to repair the infrastructure of our water and drainage systems.

Green
Awareness
Workbook

FERRIS STATE UNIVERSITY

Some exercises may require additional space.

If you require additional space for any of the exercises you may use the last four pages, which are blank

Student's Name_____

Concepts & Terminology

True & False

Circle True if the statement in the Question Text field is correct. Circle False if the statement in the Question Text field is incorrect.
Questions

1.　Energy efficiency is the relationship between the energy used and the work performed.
　　　True or False

2.　Heat load calculation includes an evaluation of a structure's ability to resist the transfer of heat.
　　　True or False

5.　The EIA provides information about post consumer waste.
　　　True or False

Multiple-choice

Circle the Correct option a, b, c, d, next to the correct answer to the question.
Questions

1. Carbon footprint is a measurement of a the _____ produced by a process or product.
　　　a) carbon monoxide
　　　b) HCFC's
　　　c) carbon dioxide
　　　d) ozone

2. Transformers that are continually plugged in but are not powering anything are an example of a
　　　a) viable load
　　　b) ghost load
　　　c) energy drain
　　　d) System drain

4. What does CBECS stand for?
　　　a) Commercial Baseline Energy Consumption Survey
　　　b) Commercial Building Efficiency Consumption Survey
　　　c) Commercial Baseline Efficiency Consumption Survey
　　　d) Commercial Building Energy Consumption Survey

5. AFUE - Annual Fuel Utilization Efficiency is;
　　　a) the percentage of a heating unit's operating fuel efficiency for the heating season.
　　　b) the annual average percentage of a heating unit's operating fuel efficiency.
　　　C) a heating unit's energy efficiency derived from the average monthly fuel usage.
　　　d) the percentage of a heating unit's operating fuel efficiency based on fuel consumption.

Student's Name_____

1. List one (1) example of an ECM (Energy Conservation Measure) for the following systems:
   Comfort heating,

   Comfort cooling,

   Lighting,

   Plumbing.

2. Describe the term "Carbon Footprint" and how it is used to describe the amount of Carbon Dioxide attached to processes, purchased commodities, and energy.

Workbook Section: HVAC

Student's Name_____

True & False
Circle True if the statement in the Question Text field is correct. Circle False if the statement in the Question Text field is incorrect.

1.   Evaporative coolers use air as the active substance to cool more air.

>   True or False

2.   Geothermal heat pumps can be coupled to solar panels.

>   True or False

3.   Solar cooling can be accomplished through the use of absorption systems.

>   True or False

4.   The majority of comfort cooling id accomplished by mechanical vapor compression systems.

>   True or False

5.   Instantaneous boilers are more efficient than standard boilers, because they do not have stand-by losses.

>   True or False

6.   Condensate drains on heating systems that do not have air conditioning, indicates that the heating system condenses water from the flue gas.

>   True or False

7.   Radiant floor heating systems generally operate at a lower water temperature as compared to a boiler systems with air handlers.

>   True or False

8.   Passive cooling uses the effect of natural heat energy.

>   True or False

9.   Condensate within a steam system becomes an energy concern when it is returned to the boiler at a lower temperature than the steam.

>   True or False

10.   Solar water heating systems use raw water in the solar collector for better heat transfer.

>   True or False

Student's Name_____

Multiple-choice
Circle the Correct option a, b, c, d, next to the correct answer to the question.

1. Misting systems cool the air by
    a) increasing air velocity   b) blowing air over a person
    c) moving air over water   d) evaporating water

2. During cooling operations, geothermal heat pumps inject
    a) heat into the geo-sink   b) water into the ground
    c) refrigerant into the dirt   d) heat into the indoor coil

3. Which of the following is a "Thermally Driven Cooling System?"
    a) absorber   b) direct expansion system
    c) boiler   d) standard air conditioning system

4. Which of the following provides cooling by removing only moisture from the air?
    a) pneumatic system   b) direct expansion system
    c) commercial air conditioning system   d) desiccant system

5. Which of the following is used as a "Thermal Storage" system?
    a) air   b) water
    c) ice   d) refrigerant

6. Bio-diesel is primarily made of which product?
    a) vegetable oil   b) refined #2 fuel oil
    c) kerosene   d) used motor oil

7. What do air to air heat pumps have that standard air conditioners do not?
    a) a compressor.
    b) an accumulator.
    c) low pressure cut-off switch to protect it in the winter.
    d) a reversing valve.

8. Condensing gas furnaces remove what substance from the flue gases?
    a) water   b) soot
    c) carbon   d) carbon dioxide

9. Modulating gas furnaces matches the amount of gases burned to the
    a) condensate in the flue gas.   b) outdoor temperature.
    c) heating need of the building.   d) customer demand setting.

10. Lost boiler heat is typically reclaimed at two locations, the stack and the:
    a) blow-off.   b) steam trap.
    c) condensate return.   d) piping run.

Student's Name_____

1. Describe how a solar hot air heating system works and draw a basic diagram.

2. Draw a solar water heating system and describe the flow of heat; from the sun and through each heat exchanger, to the heated space. Describe how the heat exchanger works.

Workbook Section: Electrical

Student's Name_____

True & False
Circle True if the statement in the Question Text field is correct. Circle False if the statement in the Question Text field is incorrect.

1.   Fuel cells can be connected together to meet the need for output voltage.

  True or False

2.   Wind energy increases at the same amount as the increase in wind speed.

  True or False

3.   Photovoltaic panels produce the same type of energy that all household appliances use.

  True or False

4.   Fluorescent lights can be dimmed like standard bulbs.

  True or False

5.   LED bulbs are estimated to last 10 years or more.

  True or False

Multiple-choice
Circle the Correct option a, b, c, d, next to the correct answer to the question.

1. Hydrogen used by fuel cells, comes from which of the following sources?

  a) wind   b) grey water

  c) atmosphere   d) fossil fuel

2. Photo Voltaic panels are best used to power which of the following?

  a) individual loads   b) subdivisions

  c) entire companies   d) cities

3. Fluorescent bulbs that are made to fit in the same location as standard bulbs are called

  a) reduced energy bulbs.   b) twisted bulbs.

  c) compact lighting systems.   d) compact fluorescent.

4. When light emitting diodes are made into bulbs for applications, many individual LEDs are placed into;

  a) panels.   b) grids.

  c) arrays.   d) banks.

Student's Name_____

1.   Describe how wind farms work to produce electricity for homes and businesses.

2.   Explain how a fuel cell works. Draw a picture and label the basic parts.

Workbook Section: Plumbing

Student's Name_____

True & False
Circle True if the statement in the Question Text field is correct. Circle False if the statement in the Question Text field is incorrect.

1. Grey water is also known as greay and gray water.
   True or False

2. Rain water is differs from grey water because it doesn't have dissolved cleaning products.
   True or False

3. Instantaneous water heaters will eventually run out of hot water.
   True or False

4. Some low volume toilets use gravity to flush.
   True or False

5. Rain water is always potable.
   True or False

Multiple-choice
Circle the Correct option a, b, c, d, next to the correct answer to the question.

1. Which of the following is usually an electrical part of a forced circulation solar domestic water heating system?

   a) pump   b) storage tank

   c) solar collector   d) piping

2. Sewer gas is prevented from entering the occupied space by which device?

   a) fixture   b) piping

   c) trap   d) valve

3. Rain water can be used for all of the following EXCEPT?

   a) irrigation   b) flushing toilets

   c) watering fruits and vegetables d) drinking

4. Which of the following needs a clean or fresh source of water?

   a) washing hands   b) flushing toilets

   c) watering fruits and vegetables d) irrigation

5. A metered flow system is used in commercial rest rooms on which of the following fixtures?

   a) hand drier   b) sink faucet

   c) toilet   d) shower head

Student's Name_____

1. Describe how a solar domestic hot water system works. Draw a picture or diagram of the system.

2. Explain how rain water might be collected, stored, and used at a later date. Draw a diagram of a rain water recovery system.